Nutrition and Skin

Apostolos Pappas
Editor

Nutrition and Skin

Lessons for Anti-Aging, Beauty and Healthy Skin

Springer

Editor
Apostolos Pappas
The Johnson & Johnson Skin Research Center
CPPW, A Division of Johnson & Johnson
Consumer Companies, Inc.
Skillman, NJ 08558, USA
apostolos_pappas@yahoo.com

ISBN 978-1-4419-7966-7 e-ISBN 978-1-4419-7967-4
DOI 10.1007/978-1-4419-7967-4
Springer New York Dordrecht Heidelberg London

Library of Congress Control Number: 2011935453

Printed on acid-free paper

Springer is part of Springer Science+Business Media (www.springer.com)

Preface

Good health has been always associated with nutrition and skin quality. It is apparent that we all desire to live longer healthy lives while maintaining a youthful appearance. A vast amount of epidemiological and clinical studies link various nutrients to health benefits in tissues and organs. Recent interest in these relations is triggering progressive reexploration by the dermatological community, particularly where connections between diet and skin have previously been dismissed. A promising volume of publications and findings now support ideas and validate theories that key nutrients are imperative for healthy skin.

Today's global economy urges food scientists and professionals to identify novel ways that can help producers reach consumers. Undoubtedly, in the world of food science the dining table is the predominant route from the food producer to the consumer. However, from any farmer who harvests flaxseeds or soybeans to every ingredient manufacturer who markets tocopherols, polyphenols, or plant extracts, it is apparent that there are many other routes to reach the consumer. The wide variety of non-food consumer products offers numerous examples.

The abundant use of vitamins and antioxidants by the cosmetic industry and their effects on skin care and dermal health has been greatly underestimated, or perhaps unseen, in the food science community, which is wholly focused on dietary use of these nutrients. Thus, not only might topical application of these products further establish the efficacy of these functional ingredients for use on skin, but their ingestion might be even more efficacious.

Current consumer trends have brought anti-aging and consumer products—from nutritional supplements to skin care—into billion dollar ranges that only drugs used to reach. All of these products are tightly connected with the health, wellness, and needs of the modern-day consumer. The main pillars of the marketing power behind these products are the pharmacological activity of "nutraceuticals."

This book serves to educate and decode the role of vitamins, essential fatty acids, and other nutraceuticals on skin health and their tremendous impact on skin health. In addition, a discussion of the potential role of functional foods is provided. Focus on skin conditions such as acne, dermatitis, dry scaly skin, or alopecia can provide comprehensive knowledge regarding the relation of nutrition and skin, as can a review of current nutritional clinical studies in dermatological research.

The contributing authors are leaders in their field who concentrate on facts and actual scientific studies. They outline the need for more studies in this new field that is so close to the heart of the consumers in our society. Indeed, the effort here is to concentrate not only on what we know but what we do not (but need to) know to meet consumers' needs. We seek to elucidate not only the potential health benefits that certain diets or nutrients bring to various tissues and organs but also the contributing effects on our skin health and visible condition. It is up to all of us—scientists, doctors, the industry, the sponsoring agencies, the government, and all the people—to find this extra time, effort, and help to address, although not life-threatening, an issue closely associated with the quality of life, health, and well-being.

Skillman, NJ Apostolos Pappas

Contents

Contributors

Michael Anthonavage Presperse, LLC, Somerset, NJ, USA

Robert J. Boyle Department of Paediatrics, Imperial College London, London, UK

Karen E. Burke Department of Dermatology, Mount Sinai Medical Center, New York, NY, USA

Maxim E. Darvin Center of Experimental and Applied Cutaneous Physiology (CCP), Department of Dermatology, Venerology and Allergology, Charité – Universitätsmedizin Berlin, Germany

Rana Mohsen Elewa Departments of Dermatology, Venereology, Allergology, and Immunology, Dessau Medical Center, Dessau, Germany

Ulrike Heinrich Institut für Experimentelle Dermatologie, Universität Witten-Herdecke, Witten-Annen, Germany

Nikiforos Kollias Johnson & Johnson Consumer Companies Inc, Skillman, NJ, USA

Juergen Lademann Center of Experimental and Applied Cutaneous Physiology (CCP), Department of Dermatology, Venerology and Allergology, Charité – Universitätsmedizin Berlin, Germany

Sampo J. Lahtinen Danisco Health & Nutrition, Kantvik, Finland

Aikaterini I. Liakou Departments of Dermatology, Venereology, Allergology, and Immunology, Dessau Medical Center, Dessau, Germany

Neil Mann School of Applied Sciences, RMIT University, Melbourne, VIC, Australia

Angus M. Moodycliffe Nestlé Research Center, Lausanne, Switzerland

Apostolos Pappas The Johnson & Johnson Skin Research Center, CPPW, A Division of Johnson & Johnson Consumer Companies, Inc., Skillman, NJ, USA

Anthony Vincent Rawlings AVR Consulting Ltd, Northwich/Cheshire, UK

Jörg Reichrath Department of Dermatology, The Saarland University Hospital, Homburg/Saar, Germany

Myriam Richelle Nestlé Research Center, Lausanne, Switzerland

Robyn Smith Baker IDI Heart and Diabetes Institute, Melbourne, VIC, Australia

Georgios N. Stamatas Johnson & Johnson Consumer France, Issy-les-Moulineaux, France

Mimi L.K. Tang Department of Allergy and Immunology, Royal Children's Hospital, Melbourne, VIC, Australia

Murdoch Children's Research Institute, Melbourne, VIC, Australia

Department of Paediatrics, University of Melbourne, Melbourne, VIC, Australia

Sagar K. Thakkar Nestlé Research Center, Lausanne, Switzerland

Michael J. Theodorakis University of Athens Medical School, Athens, Greece

Léa Trémezaygues Department of Dermatology, The Saarland University Hospital, Homburg/Saar, Germany

Petra Winkler J&J, Medical Affairs, EMEA, Neuss, Germany

Christos C. Zouboulis Departments of Dermatology, Venereology, Allergology, and Immunology, Dessau Medical Center, Dessau, Germany

Chapter 1
Introduction and Overview

Apostolos Pappas

Throughout history, our traditions, anecdotal evidence, epidemiological studies, and most recently clinical studies have reflected the idea that nutrition is associated with health, beauty, and graceful aging. Multiple pathways and cofactors are implicated in skin biology, and certain common skin conditions have been shown to be critically affected by nutritional patterns and habits.

Linking food and good nutrition to overall health and appearance is important to the modern consumer. It seems that everyone, from the youngest to the oldest consumer, is very much aware of his or her appearance. It is instinctual that food and wellness are closely associated, and skin is the only organ that is tightly associated with the desirable appearance. Skin is the largest organ of the human body and plays a role in thermoregulation, protection, metabolism, and sensation. Various nutrients are fundamental for normal skin functions, and their presence and function still intrigue many scientists.

The early part this book explores the relations of the most important and essential nutrients with skin. Why are they the most important and essential (together)? These two terms have harmoniously coinhabited books of nutritional research. The term essential has been used mainly to denote nutrients that the human body cannot synthesize and relies on dietary sources to provide; they include, for example, vitamins, essential fatty acids, and amino acids as well as many bioactive agents with beneficial pharmacological activity that belong to the wide category of nutraceuticals.

Zouboulis and Elewa recap the essentiality of vitamin A for normal differentiation and maintenance of epithelial tissues in skin and mucous membranes. They comprehensively review the multiple activities of retinoids (metabolites of vitamin A)

A. Pappas (✉)
The Johnson & Johnson Skin Research Center, CPPW, A Division of Johnson & Johnson Consumer Companies, Inc., Grandview Road 199, Skillman, NJ 08558, USA
e-mail: apostolos_pappas@yahoo.com

A. Pappas (ed.), *Nutrition and Skin: Lessons for Anti-Aging, Beauty and Healthy Skin*,
DOI 10.1007/978-1-4419-7967-4_1, © Springer Science+Business Media, LLC 2011

that are used as therapeutic agents for various skin diseases. They discuss the essential aspects, including the profound management of these potent agents.

Reichrath and Trémezaygues undertook the enormous task of discussing the multifaceted roles of vitamin D and its biologically important metabolites. The studies are numerous, but the authors have carefully reviewed them for this chapter, summarizing the positive effects of vitamin D and its analogues in a variety of skin diseases.

Burke eloquently reviews the photoprotection that vitamins C and E offer against induced photodamage. Their activities are conversed in full spectrum, covering both topical and oral use. Their synergies with other antioxidants are discussed in detail.

Richelle's team powerfully compiled information on the most studied antioxidants in skin research to date, the carotenoids. They have accurately and carefully elaborated on important issues such as their bioavailability and their protective role against solar ultraviolet damage; a role that may significantly complement the use of sunscreens and overall contribute extensively to skin health.

Lademann's team complements successfully the rest of the vast range of antioxidant activities and their observed benefits to skin, revealing in addition ways to assess their important activities. The nature of free radical scavengers is discussed in detail, especially in regard to how they reflect both the lifestyle and the physical condition of people.

Petra Winkler excels in her difficult task to review and focus on the knowledge currently available on several minerals and their impact on the skin. The focus on the deficiencies of certain minerals is discussed together with their association with many skin diseases.

Finally, Boyle and colleagues offer a powerful message on the beneficial and perplexing relation of probiotics and skin with a complete review of the published clinical research. Many manuscripts loosely associate our microbiota to health. Boyle's team focus on the facts—using direct evidence of their beneficial effects on skin. Also, a special focus on eczema prevention and an additional call for further studies is outlined.

The latter part the book focuses on the clinical crossover between nutrition and dermatology. There is a strong need to expand this knowledge and an urgent wish to see it enriched and more powerful in the near future.

The first topic discussed in this portion of the book is how nutrition may play a role in acne. Many nutritional factors are discussed especially essential nutrients and dairy products. Also addressed is the need for more proper clinical studies that would help elucidate the relation between diet and acne. So far, only studies that examined the high glycemic index and load have been able to establish a valid relation between diet and acne. Thus, in the next chapter Larsen (Smith) discusses her pioneering studies on the relation of the glycemic load and acne symptoms through improvements in insulin metabolism.

Consequently, Rawlings distills years of experience into a chapter that addresses atopic dermatitis as it relates to nutrition, with a focus on essential fatty acid metabolism. Stratum corneum health, especially its skin barrier function, is discussed in relation to its deficiencies due to essential fatty acids. These results and intervention studies with these fatty acids are summarized.

Anthonavage identifies, with detail and depth, research areas unchartered by nutritionists and food scientists with respect to hair biology. As the fundamental element of beauty and noticed appearance, hair encapsulates a range of factors and information that relates to nutritional status and metabolic activities.

Stamatas and Kollias, in a comprehensive summary, capture the enormous potential that are available to scientists and health professionals in the form of imaging and spectroscopic techniques. These innovations can be employed to assess the presence and amount of certain nutrients directly in the skin.

The circle closes with another chapter by Zouboulis's team, who undertook the extensive task of summarizing the nutritional clinical studies in dermatology. Nutritional supplementation, caloric content and composition, nutritional imbalances, hyperinsulinemia, and important nutrients are summarized in relation to characteristic skin pathologies, skin metabolism, skin physiology, and skin properties.

My hope is that this publication bridges a gap in scientific knowledge and encourages food science and nutritional professionals to view dermatological research with a new perspective. It is also hoped that it will inspire new research and product development to help with more efficient decisions by today's highly anticipating consumer.

May this book be an asset not only for the food scientist, nutritionist, and dermatologist but also any scientist who has a pathos for food, nutrients, and metabolic research. I hope it will be the cornerstone for more books to come on this subject and, more importantly, more studies to be done in this new field. Furthermore, it is my hope that it will bridge the gap between the two disciplines.

Part I
Nutrients and Skin

Chapter 2
Vitamin A and the Skin

Rana Mohsen Elewa and Christos C. Zouboulis

Core Messages

- Vitamin A (vit. A) is essential for normal differentiation and maintenance of epithelial tissues in skin and mucous membranes, vision (retinaldehyde), reproduction (retinol), and embryonic morphogenesis.
- Retinoids (compounds with biological activities similar to those of vit. A) are used as therapeutic agents for hyperkeratotic and parakeratotic skin diseases, acne and acne-related disorders, and hand eczema. They are also used as prophylaxis for epithelial skin tumors in immune-suppressed patients and as therapy for non-melanoma skin cancers and cutaneous T-cell lymphoma.
- Regular monitoring is essential for the avoidance and management of a wide range of adverse effects.

2.1 Introduction

Retinoids refer to compounds that have biological activities similar to those of naturally occurring vitamin A (vit. A) but not necessarily the same chemical structure (Zouboulis and Orfanos 2000). The definition of retinoids does not include a structural homology to vit. A. Retinoids are used as therapy and prophylaxis systemically or as local applications for various skin diseases and tumors. They are also used in the cosmetic field for acne, seborrhea, psoriasis, epithelial tumors, and hand eczema. Retinoids perform their actions through binding to

---

C.C. Zouboulis (✉)
Departments of Dermatology, Venereology, Allergology, and Immunology,
Dessau Medical Center, Auenweg 38, 06847 Dessau, Germany
e-mail: christos.zouboulis@klinikum-dessau.de

A. Pappas (ed.), *Nutrition and Skin: Lessons for Anti-Aging, Beauty and Healthy Skin*,
DOI 10.1007/978-1-4419-7967-4_2, © Springer Science+Business Media, LLC 2011

retinoid receptors (RAR and RXR), members of the hormone nuclear receptor subfamily. In addition to its toxicity, a deficiency of vit. A causes skin manifestations. Owing to the wide range of adverse effects of retinoids, precautions and regular monitoring are essential for long-term therapy.

2.2 Naturally Occurring Retinoids

Natural retinoids include vit. A (retinol) and its metabolic derivatives retinaldehyde and retinoic acid (Zouboulis and Orfanos 2000). The normal concentration of vit. A in plasma is 0.35–0.75 μg/mL. Retinoic acid is produced by in vivo oxidation of retinol. Its two isoforms are all-trans retinoic acid and 13-*cis* retinoic acid, with normal plasma concentrations of 0.55–1.20 and 0.80–2.40 ng/mL, respectively. Retinoic acid can fully substitute for retinol, except for maintaining reproduction. Daily requirement of vit. A is 0.8–1.0 mg (2,400–3,000 IU), which can be found in ten medium-sized eggs or 100 g of butter. Hypervitaminosis occurs with intake of more than 18,000–60,000 IU vit. A per day for children and 50,000–1,00,000 IU for adults. Higher uptakes usually do not result in elevated retinol levels, however the stored retinol esters level are increased wherever retinol esters are stored.

2.3 Synthetic Retinoids

The synthetic retinoids (Table 2.1) are either chemical modifications of naturally occurring vit. A or chemically different compounds with the capacity to bind or antagonize retinoid nuclear receptor proteins (Zouboulis and Orfanos 2000). Nonaromatic, monoaromatic, and polyaromatic compounds have been developed through chemical modification. Modification of the polyene chain diminishes bioactivity; modification or esterification of the carboxylic end diminishes the toxicity while maintaining the activity; and ring substitution diminishes the toxicity and markedly increases the activity.

2.4 Absorption, Distribution, and Metabolism

The oral bioavailability of retinoids can be increased, especially by fatty acids, which prevent the binding of retinoids with albumin and hence improve the clinical effect (Avis et al. 1995). The metabolism of retinoids generally occurs in the liver. It involves oxidation and chain shortening to produce biologically inactive metabolites. Retinoids are excreted through feces and urine.

Isotretinoin is detectable after 30 min in blood, and maximum concentrations are reached 2–4 h after oral intake (Gancevicience and Zouboulis 2007). The half-life elimination rate of isotretinoin ranges from 7 to 37 h, and that of its known

Table 2.1 Synthetic retinoids in current use

Chemical structure	Generic name	Initial trade name	Approved medical use
First generation			
All-trans retinoic acid	Tretinoin	Retin-A	Topical treatment of acne vulgaris, photodamage
		Vesanoid	Systemic treatment of acute promyelocytic leukemia
9-*cis* retinoic acid	Alitretinoin	Toctino	Systemic treatment of therapy-resistant hand eczema
		Panretin	Topical treatment of AIDS-associated Kaposi sarcoma
13-*cis* retinoic acid	Isotretinoin	Accutane	Systemic treatment of severe recalcitrant acne and acne-related dermatoses
		Isotrex	Topical treatment of acne
Second generation			
Monoaromatic compound	Etretinate	Tigason	Systemic treatment of psoriasis
Etretinate derivative	Motretinide	Tasmaderm	Topical treatment of acne vulgaris
Free acid metabolite of etretinate	Acitretin	Neotigason	Systemic treatment of psoriasis and severe disorders of keratinization
Third generation			
Polyaromatic retinoids (retinoids)	Adapalene	Differin	Topical treatment of acne
	Tazarotene	Tazorac	Topical treatment of plaque-type psoriasis, acne vulgaris, and photodamage
	Bexarotene	Targretin	Systemic treatment of cutaneous T-cell lymphoma, breast cancer, and Kaposi sarcoma
			Topical treatment of cutaneous T-cell lymphoma

metabolites is 11–50 h (Benifla et al. 1995). The major metabolites of isotretinoin in blood are 4-oxo- and 4-hydroxy-isotretinoin; several glucuronide conjugates are detectable in bile (Vane et al. 1990). Because there is interconversion between the two isomers, isotretinoin and tretinoin, in vivo, about 10–30% of the drug is metabolized via tretinoin. Isotretinoin is excreted in feces after conjugation or in urine after metabolization. In contrast to vit. A, there is neither liver nor adipose tissue storage. More than 99% of isotretinoin in plasma is bound to plasma proteins, mainly albumin (Rollman and Vahlquist 1986). Serum albumin has a critical function as a retinoid-binding protein in reducing the concentration of active retinoids and restricting the biological effects on sebaceous gland cells (Tsukada et al. 2002). After discontinuation of therapy, isotretinoin disappears from serum and skin within 2–4 weeks. It seems likely that isotretinoin therapy interferes with the endogenous metabolism of vit. A in the skin because vit. A levels increased by about 50% and dehydrovitamin A levels decreased by around 80% in some patients (Orfanos and Zouboulis 1998).

Acitretin is eliminated more rapidly than etretinate. Etretinate is highly lipophilic, binds strongly to albumin, is stored in adipose tissue, and is released slowly (half-life of 120 days), whereas the half-life of acitretin is only 2 days. Reesterification of acitretin to etretinate occurs only in cases of alcohol consumption (Wiegand and Chou 1998).

Bexarotene in plasma is 99% bound to plasma proteins. It is excreted via the hepatoiliary system, and its terminal half-life is 7–9 h (Ethan Quan and Wolverton 2003).

2.5 Mechanism of Action

2.5.1 Retinoid Receptors and Gene Regulation

Retinoids enter the cell by non-receptor-mediated endocytosis, interact with cytosolic proteins, and finally bind to nuclear receptors. The retinoid nuclear receptors are members of the steroid thyroid hormone receptor superfamily. Retinoid A receptors (RAR) bind all-trans retinoic acid and 9-*cis* retinoic acid with high affinity, whereas they barely bind to 13-*cis* retinoic acid. Retinoid X receptors (RXR) also bind 9-*cis* retinoic acid and are selective for bexarotene, which is a specific RXR ligand (rexinoid). 14-Hydroxy-retro-retinol does not bind or activate retinoid receptors, whereas acitretin does not bind but does activate RAR.

The RAR and RXR families each include three members: α, β, γ. Each receptor is mapped on a different chromosome. High expression of RAR-γ and RXR-α was detected in healthy and psoriatic epidermis, as well as in sebaceous gland cells.

Retinoid receptors target and regulate the genes that have retinoid-responsive elements (RARE and RXRE) in their promoter regions. The retinoid receptor–gene interaction occurs because RAR genes have retinoid-responsive elements, which allow a positive feedback mechanism. On the other hand, the retinoid-binding proteins may antagonize the retinoid interaction with their nuclear receptors. Specific retinoid effects can occur via interaction of retinoid receptors with other signal transduction mechanisms.

2.5.2 *Effect on Epidermal Growth and Differentiation*

Retinoids promote cell proliferation in normal epithelia, whereas they normalize it in hyperproliferative conditions. Retinoids induce and modulate the expression of growth factors and their receptors. Keratinocyte proliferation by retinoids is mediated through induction of cyclic adenosine monophosphate (cAMP), epidermal growth factor (EGF) receptor binding, protein kinase C (PKC), and transforming growth factor-α (TGFα). Retinoids down-regulate cell growth which is mediated by a TGFβ2-regulated inhibition of the EGF binding to its receptor. Retinoids have also a keratolytic effect through shifting the terminal keratinocyte differentiation toward a nonkeratinizing mucosa such as epithelium. As for differentiation, retinoic acids down-regulate most of the markers of terminal differentiation in vitro (loricrin, transglutaminase, involucrin, filaggrin, keratins 1 and 10), whereas keratins 19 and 13, the markers of nonstratified and wet epithelia, are up-regulated. Adapalene and retinoic acid restore the architecture of epidermis and antagonize hyperkeratosis. RAR-α receptor agonists promote differentiation in T47D breast carcinoma cells in vitro.

2.5.3 *Effects on Sebaceous Gland Activity*

See the section Seborrhea, Acne, and Acneiform Disorders later in the chapter.

2.5.4 *Immunomodulatory and Antiinflammatory Properties*

Isotretinoin, etretinate, and acitretin were proved to inhibit angiogenesis both in vitro and in vivo (in mice with T47D cell-induced tumors) most probably through RAR-α. Retinoids can stimulate humoral and cellular immunity through enhancing antibody production, increasing blood T-helper cells, and preventing Langerhans cells depletion from epidermis by ultraviolet (UV) light and in vitro increasing cell surface antigens of T and natural killer cells. Antiinflammatory activity of retinoids has also been shown. Inhibition of neutrophil migration in psoriatic skin, inhibiting LTB4-induced migration of neutrophils (topical isotretinoin more than tretinoin or retinoids), and inhibition of nitric oxide and tumor necrosis factor-α (TNFα) production have been reported.

2.6 Therapeutic Uses

2.6.1 Psoriasis and Related Disorders

The orally administered aromatic retinoids (etretinate and acitretin) are used to treat psoriasis, both initially and for maintenance. Retinoids have a synergistic effect with other psoriasis treatment modalities. They are considered the drug of choice in cases of pustular psoriasis and palmoplantar psoriasis. Retinoids were found effective in the juvenile types of pityriasis rubra pilaris. Retinoids act on both the epidermal and dermal levels to exhibit their antipsoriatic action. They reduce proliferation, enhance differentiation, regulate the desquamation of the corneocytes, modulate lymphocyte function, and inhibit neutrophil migration.

The daily dose of acitretin is 0.5 mg/kg body weight divided into two administrations to avoid serum peaks and complications (Table 2.2). Taking acitretin with meals that include some fat increases the blood absorption two- to fivefold. Retinoids being metabolized in the liver interact with ketoconazole and phenytoin but not with oral contraceptives.

2.6.2 Disorders of Keratinization

Etretinate and acetretin are superior to isotretinoin because of the latter's sebum-drying property. The severity of Darier's disease, ichthyosis vulgaris, congenital ichthyosis, and palmoplantar keratodermas can be successfully controlled with retinoids. Usually the treatment is prescribed with a low dose (0.2–0.5 mg/kg/day); a lifelong maintenance dose is required with long-term contraception and control of

Table 2.2 Doses of acitretin and therapeutic effect

Disease	Dose	Duration of therapy
Pustular psoriasis	High initial dose (0.5–1.0 mg/kg) reduced to 0.20–0.25 mg/kg over 3–6 months	Maintenance 6–12 months
Erythrodermic psoriasis	Low initial dose (0.20–0.25 mg/kg) increased to 0.5–0.6 mg/kg over 3 months	Maintenance 6 months
Plaque-type psoriasis		
Monotherapy or combined with anthraline or topical steroids	0.3–1.0 mg/kg/day	4–12 weeks
With UVB	0.2–0.5 mg/kg/day	6 weeks
With PUVA	0.2–0.5 mg/kg/day	4–6 weeks

UVB ultraviolet B radiation; *PUVA* psoralen + ultraviolet A radiation bone density to avoid bone toxicity. Interestingly, other keratinization disorders such as pachyonychia congenita, inflammatory linear verrucous epidermal (ILVEN) nevus, Netherton's syndrome, and monilethrix do not respond to retinoids,

2.6.3 Seborrhea, Acne, and Acneiform Disorders

Isotretinoin proved to be the most effective sebostatic retinoid both in vivo and in cell cultures; and it is the best retinoid for treating severe acne (Zouboulis and Orfanos 2000). Although it shows low binding affinity for intracellular retinoid-binding proteins and for RAR and RXR, isotretinoin has strong sebostatic activity. It undergoes a specific and selective intracellular isomerization process into tretinoin, which in turn binds to RARs. The superior sebostatic effect of isotretinoin over tretinoin (its metabolite) is attributed to the delayed initiation of inactivation of isotretinoin; incubation of sebocytes with tretinoin leads to rapid enhancement of cellular retinoic acid-binding protein levels, which promotes the metabolism by cytochrome P 450 enzymes (Tsukada et al. 2000). It was also found that isotretinoin inhibits 3α-hydroxysteroid oxidation, leading to decreased levels of dihydrotestosterone and androstanedione, which may contribute to the sebosuppressive effect (Guy et al. 1996; Karlsson et al. 2003). Lipogenesis is also reduced by TGFβ2 and TGFβ3, which are rapidly and transiently expressed as a response to retinoid administration (Downie et al. 2002). Isotretinoin decreases sebum production, decreases the number of proliferating sebocytes and the size of the sebaceous gland, inhibits sebocyte differentiation in vivo and in vitro (Orfanos et al. 1997; Zouboulis et al. 2000; Ganceviciene and Zouboulis 2007), and directly suppresses abnormal desquamation of sebaceous follicles. Isotretinoin thus alters the follicular microenvironment via its sebostatic effect and hence markedly reduces the *Propionibacterium acnes* count (King et al. 1982). It also has immunosuppressive and antifibrosis effects when tested on renal allografts (Adams et al. 2005). Through RAR-independent mechanisms, it was associated with cell cycle arrest, induction of apoptosis, decreased proliferation, decreased lipogenesis rate, and decreased DNA synthesis (Zouboulis et al. 1993; Nelson et al. 2006; Zouboulis 2006a). Isotretinoin acts in a receptor-independent manner by influencing cellular signaling pathways through direct protein interactions or by enzyme inhibition (Imam et al. 2001).

Isotretinoin reduces monocyte and neutrophil chemotaxis and their migration to the epidermis, minimizing the excessive inflammation that causes scarring (Zouboulis 2006a). The matrix metalloproteinases (MMPs) were found to be elevated in acne lesions, raising a possibility of involvement in acne pathophysiology through mediation of inflammation and collagen degradation. Isotretinoin induced reduction of Pro-MMP-9 and MMP-13 (Papakonstantinou et al. 2005). Isotretinoin has a strong influence on sebaceous lipid composition, as it decreases wax esters, triglyceride fractions, and squalene; it relatively increases the cholesterol level and the levels of free sterols and total ceramides (Orfanos and Zouboulis 1998).

2.6.3.1 Dosing, Therapeutic Effect, and Monitoring

The required dose is 0.5 mg/kg/day, an initially high dose for 3 months; maintenance requires a lower dose. A cumulative dose of more than 150 mg/kg administered over 6–12 months has been considered necessary to ensure a long-lasting remission (Zouboulis and Piquero-Martin 2003). Owing to relevant recurrence rates, current treatment concepts individualize dosage and duration (Zouboulis 2006b). Longer treatment duration might be needed in patients with extrafacial lesions, low-dose therapy, or severe acne. Contraception is essential during treatment with isotretinoin, and using an antiandrogen-containing contraception is of a great value (Zouboulis and Rabe 2010).

The European directive recommendations for the use of isotretinoin for treatment of acne note that treatment should start at 0.5 mg/kg, and that it should be used only for severe acne (nodular or conglobata) that is not responding to antibiotics or topical therapy; it should not be used as a first-line treatment. It is not recommended in children under 12 years of age. The liver enzymes and serum lipids should be checked before therapy, 1 month after its initiation, and every 3 months after that. All forms of peeling and wax depilation should be avoided during therapy and 6 months afterward. The pregnancy-preventing program for female patients during their childbearing period includes a medically supervised pregnancy test before, during, and 5 weeks after therapy begins. The test should be repeated monthly, and double contraception should be used. Only 30 days of oral isotretinoin can be supplied to female patients at a time (Layton et al. 2006).

2.6.3.2 Retinoid Local Therapy in Acne

Tretinoin, isotretinoin, motretinide, adapalene, and tazarotene are used for the local therapy of acne. Retinaldhyde, retinol, and retinyl esters, however, are used in cosmetic preparations. Early use of retinoids is recommended (Gollnick et al. 2003). Topical retinoids were found to perform their therapeutic action by increasing follicular epithelium turnover, reversing abnormal desquamation of the sebaceous duct. Hence the mature comedones are expelled and formation of microcomedones is inhibited; moreover, the aerobic follicular environment no longer favors the growth of *Propionibacterium acnes* (Lavker et al. 1992; Thielitz et al. 2007). Retinoids also show immunomodulatory activities (Bikowski 2005; Jones 2005).

Tretinoin and isotretinoin can be used alone or in combination with topical erythromycin, clindamycin, or benzoyl peroxide in different concentrations (Mills and Kligman 1978;Korting and Braun-Falco 1989; Marazzi et al. 2002).

Adapalene 0.1% was proved to have the same efficacy as tretinoin 0.025% but with more rapid onset of action (Cunliffe et al. 1998). Adapalene was also used in combination with clindamycin 1%, benzoylperoxide, or both (Zhang et al. 2004; Thiboutot et al. 2007; Del Rosso 2007). Addition of adapalene local therapy to the systemic doxycycline 100 mg/day significantly raised its efficacy (Thiboutot et al. 2005).

Tazarotene, which has been approved for topical acne treatment only in the United States, showed a stronger reduction in disease severity than adapalene but with slightly higher irritation (Webster et al. 2002). Also, as for other topical retinoids, the combination with clindamycin or benzoyl peroxide (BPO) showed better efficacy (Tanghetti et al. 2006).

Maintenance therapy is required for a chronic disease such as acne. The use of tazarotene gel alone proved to be as effective as a combination of tazarotene and minocycline 100 mg/day but without the side effects of the systemic antibiotic (Zhang et al. 2004). Topical adapalene is also adequate for maintenance treatment.

2.6.4 Retinoids in Skin Cancer

2.6.4.1 Prevention of Keratinocyte Skin Cancers

Retinoids are used for chemoprevention and chemosuppression in many diseases and syndromes with high susceptibility of nonmelanoma skin cancer development (Abdel Naser and Zouboulis 2010; Nguyen and Wolverton 2000). Retinoids regulate MMPs, TGFβ, cyclin-dependent kinase 1, P16, and P21; hence, they are capable of regulating tumor stroma production and control tumor progression and invasion (Ayer et al. 1995; Hassig et al. 1997). RAR-related antiproliferative and proapoptotic signals may also be involved (Sun et al. 2000). Inhibition of the AP-1 complex, suppression of cyclooxygenase-2 (COX-2) expression, and inhibition of prostaglandin E_2 (PGE2) synthesis are all involved in the reduction of cell proliferation (Fanjul et al. 1994; Kanekura et al. 2000). Specific indications of using retinoids in tumor prevention and suppression include patients with xeroderma pigmentosum, epidermodysplasia verruciformis, basal cell nevus syndrome (Gorlin-Goltz), or Bazex and Rhombo syndromes as well as individuals under immune suppression for organ transplantation (Shuttleworth et al. 1988; Otley et al. 2006).

Isotretinoin, because of its short half life, is the treatment of choice in young patients with xeroderma pigmentosum. Acitretin and etretinate are used in older patients and organ transplantation recipients who exhibited five or more low-risk squamous cell carcinomas (SCCs) per year, two or more high-risk SCCs per year, or systemic SCC (Kovach et al. 2006). Bexarotene (300 mg/m^2/day) is the choice for cutaneous T-cell lymphoma (CTCL) (Duvic et al. 2001a, b). The treatment should be continued indefinitely, as discontinuation is associated with a rebound increase in skin cancer frequency (Kraemer et al. 1988; Goldberg et al. 1989). Large doses are required for chemosuppressive and preventive effects (isotretinoin 2 mg/kg/day, maintenance dose 1.5 mg/kg/day) (Lippman et al. 1987; Shuttleworth et al. 1988); therefore, close monitoring is essential. A combination of topical tretinoin and a low-dose etretinate (10 mg/day) may reduce the adverse effects of oral medication (Rook et al. 1995). The indications must be carefully assessed to outweigh the wide range of complications and adverse effects. On the other hand, isotretinoin plus interferon-α for adjuvant therapy of aggressive skin SCC showed high rates of tumor recurrence and second primary tumors (Brewster et al. 2007).

2.6.4.2 Therapy of Other Skin Cancers

Melanoma was proved to be nonsensitive to retinoids. Bexarotene gel was introduced as a monotherapy for treatment of CTCL in early stages (IA, IB, IIA), which were refractory or intolerant to at least two other treatment modalities for more than 6 months (Zouboulis 2001). Bexarotene was found to induce a 50% overall inhibitory response in patients with refractory or persistent CTCL when administered either orally or topically with minimal toxicity. The use of oral bexarotene first to reduce dermal infiltrates prior to DAB(389)IL-2 (denileukin diftitox) administration might reduce subsequent side effects imparted by this therapy (Duvic 2000).

Alitretinoin, as a 0.1% gel, was introduced as an adjuvant topical regimen for Kaposi sarcoma associated with acquired immunodeficiency syndrome (AIDS) (Walmsley et al. 1999).

2.6.5 Chronic Hand Eczema

Eczema of the hands is a relatively common disease, seen in 6–8% of the population. Alitretinoin given at well-tolerated doses induces substantial clearing of chronic hand dermatitis in patients refractory to conventional topical therapy (Thomas et al. 2004).

Oral alitretinoin induced clinically significant responses in a high percentage of patients with moderate or severe chronic hand dermatitis refractory to standard topical therapy. Complete or near-complete disappearance of disease signs and symptoms was reported for 53% of patients treated with the highest alitretinoin dose (40 mg/day). The response was reported with all types of chronic hand eczema. Treatment is generally tolerated, with the typical retinoid adverse effects at the highest dose (headache, flushing, elevated serum fat levels) (Ruzicka et al. 2004). The mechanism of action of alitretinoin for chronic hand eczema may be explained by the fact that alitretinoin is a panretinoid agonist activating RXRs and RARs, mediating the stimulation of Th2 immune function (Bollag and Ott 1999; Stephensen et al. 2002). Bexarotene, applied topically in 1% concentration also proved to be effective for treatment of chronic hand eczema (Hanifi et al. 2004).

2.7 Adverse Reactions and Tolerability

2.7.1 Mucocutaneous Complications

The most common complication associated with retinoids is skin and mucous membrane dryness, including cheilitis, which occurs in about 90% of cases; it is an indication of good absorption. Mucocutaneous xerosis (30%), nosebleeds (15%), mild hair loss, augmented skin fragility, and palmoplantar desquamation (80%) are additional adverse effects. The use of skin emollients, artificial eye tears, Vaseline for the inner part of the nose, and moisturizers for the lips are a necessity from the start of the treatment to avoid the dry skin complications. In the case of extensive dryness, reducing the dose by 25% can be of great help (Otley et al. 2006). Capillary leak syndrome, a rare complication that leads to face or generalized edema, was also reported (Estival et al. 2004; Scheinfeld and Bangalore 2006).

2.7.2 *Ocular and Neurological Complications*

Decreased tear production and decreased lipid content of the tear film lead to dryness of the eyes, and keratitis and corneal erosions can occur. The use of contact lenses is contraindicated, and artificial tears are mandatory. Other complications include blurred vision, decreased vision, photophobia, decreased dark adaptation, papilledema, corneal opacities, and retinal dysfunction. Blepharoconjunctivitis, which may be complicated with *Staphylococcus* infection, occurs in 20–50% of cases.

Pseudotumor cerebri (benign intracranial hypertension) is the most important neurological side effect. It manifests as headache accompanied by nausea, vomiting, and visual changes due to papilledema. Half of the reported patients were taking tetracycline or minocycline concomitantly with isotretinoin. Although depression, irritability, and suicidal intentions have been reported, a causal relation has not been demonstrated (Gold et al. 1989; Enders and Enders 2003; Scheinfeld and Bangalore 2006).

The most frequent central nervous system adverse effect associated with oral isotretinoin is headache, either as an independent adverse effect or as part of benign intracranial hypertension. Isolated cases of stiff-person-like syndrome, epileptic seizures, and generalized muscle stiffness syndrome possibly or probably related to oral treatment with isotretinoin have been reported (Chroni et al. 2010).

2.7.3 *Serum Lipids, Gastrointestinal Side Effects, Liver Function, and Endocrine Adverse Effects*

Increased serum lipids—hypertriglyceridemia (20–40%) being more common than hypercholesterolemia (10–30%) (Olsen et al. 1989; Barth et al. 1993)—with decreased high-density lipoprotein and increased low-density lipoproteinemia as well as mild transaminase increase are characteristic metabolic side effects of retinoids. Lipid profile changes are more likely to occur in patients with predisposing factors as obesity, familial hyperlipidemia, nicotine abuse, and diabetes as well as in patients using β-blockers and contraceptive pills (Orfanos et al. 1997). These changes are proportionate to the dose and reversible within 1–2 months (Gupta et al. 1989). Nausea, diarrhea, and abdominal pain may occur. Hyperlipidemia can be avoided to an extent by ingesting a low-fat diet and by hypolipidemic drugs when needed. Chronic liver toxicity is rather rare, although acutely elevated liver enzymes is not uncommon mostly occurring with etretinate (Zane et al. 2006).

Monitoring of the liver enzymes and blood lipids according to the "European directive for prescribing systemic isotretinoin for acne vulgaris" must be done before the beginning of the therapy, after 1 month, and subsequently every 3 months. The concomitant use of methotrexate (in cases of acitretin treatment of psoriasis) or other drugs that affect the liver must be avoided. If the liver enzymes are elevated one- to threefold over normal levels, the therapy should be continued with 50% of the dose and the enzymes reevaluated in 2 weeks. Discontinuing the therapy is advised when liver function tests are three times normal, followed by a regular enzyme checkup every 2 weeks. Reintroduction is recommended after enzyme level normalization with 25% of the original dose (Otley et al. 2006).

Bexarotene causes significant central hypothyroidism (Haugen 2009) and hyperlipidemia in most patients managed with thyroid replacement and hypolipidemic drugs, respectively (Abbott et al. 2009). Hypothyroidism is in part due to increased peripheral thyroid hormone metabolism (Smit et al. 2007).

2.7.4 Long-Term Toxicity: Bone Changes

The long-term bone toxicity occurs mostly with vitamin A chronic toxicity. The changes include hyperostosis and periostosis, demineralization, thinning of the bones, and premature epiphysial closure (Biesalski 1989). Radiography of significantly symptomatic joints is recommended with long-term therapy. Yearly radiography of the ankle or thoracic spine is optional (Otley et al. 2006; Scheinfeld and Bangalore 2006).

2.7.5 Arthralgias and Myalgias

Diffuse idiopathic skeletal hyperostosis-like hyperostosis, lesions mimicking seronegative spondyloarthropathy, arthralgia, myalgia, stiffness, true myopathy, muscular damage, rhabdomyolysis, and museuloskeletal pain can occur under retinoid treatment, especially isotretinoin. Creatine phosphokinase, a specific marker of muscle destruction, has been found to be elevated, occasionally by up to 100 times the normal value (with or without muscular symptoms and signs) in a variable percentage of patients receiving isotretinoin treatment and particularly in those undergoing vigorous physical exercise. Oral acitretin has been found to cause peripheral nerve dysfunction, particularly of sensory fibers, which in rare cases leads to clinically evident sensory disturbances. Less clear is the causal relation between acitretin and benign intracranial hypertension or myopathy, whereas an isolated case of cranial nerve IV (oculomotor) palsy and another case of thrombotic stroke during treatment with oral acitretin have been reported (Chroni et al. 2010).

Systemic diseases with involvement of nervous and/or muscle tissue and neuromuscular disorders should be regarded as exclusion criteria for initiation of oral retinoid therapy. Additionally, intense physical exercise and concurrent treatment with neurotoxic or myotoxic drugs should be avoided during treatment with oral retinoids (Chroni et al. 2010). When arthralgias and myalgias occur, it is recommended that the dose be decreased by 25% until the symptoms resolve (Otley et al. 2006).

2.7.6 *Teratogenicity*

Retinoids freely cross the placenta, causing severe fatal fetal malformations including craniofacial deformities, bone and cardiovascular abnormalities, and endocrine malfunctions. All systemic retinoid treatments are contraindicated during pregnancy (U.S. Food and Drug Administration category X). Exclusion of pregnancy before starting therapy and contraception during the therapy is mandatory. Contraception should be continued for 1 month after cessation of the retinoid therapy (isotretinoin and bexarotene) and for 3 years after acitretin. Topical retinoids are also contraindicated.

The pregnancy-preventing program for female patients during their child-bearing period is strictly applied to isotretinoin therapy. This includes a medically supervised pregnancy test before, during, and 5 weeks after therapy begins. The test should be repeated monthly, and double contraception should be used. For female patients, only 30 days of oral isotretinoin can be supplied at a time (Layton et al. 2006).

References

Abbott RA, Whittaker SJ, Morris SL, Russell-Jones R, Hung T, Bashir SJ, et al. Bexarotene therapy for mycosis fungoides and Sézary syndrome. Br J Dermatol. 2009;160:1299–307.

Abdel Naser MB, Zouboulis CC. Retinoids in the management of non melanoma skin cancer. In: Jemec GBE, Kemény L, Mieh D, editors. Non-surgical treatment of keratinocyte skin cancer. Berlin/Heidelberg: Springer; 2010. p. 187–95.

Adams J, Kiss E, Arroyo AB, Bonrouhi M, Sun Q, Li Z, Gretz N, Schnitger A, Zouboulis CC, Wiesel M, Wagner J, Nelson PJ, Gröne HJ. 13-cis retinoic acid inhibits development and progression of chronic allograft nephropathy. Am J Pathol. 2005;167:285–98.

Avis I, Mathias A, Unsworth EJ, Miller MJ, Cuttitta F, Mulshine JL, et al. Analysis of small cell lung cancer cell growth inhibition by 13-cis-retinoic acid: importance of bioavailability. Cell Growth Differ. 1995;6:485–92.

Ayer DE, Lawrence QA, Eisenman RN. Mad-Max transcriptional repression is mediated by ternary complex formation with mammalian homologs of yeast repressor Sin3. Cell. 1995;80:767–76.

Barth JH, MacDonald-Hull SP, Mark J, et al. Isotretinoin therapy for acne vulgaris: a re-evaluation of the need for measurements of plasma lipids and liver function tests. Br J Dermatol. 1993;129:704–7.

Benifla JL, Ville Y, Imbert MC, Frydman R, Thomas A, Pons JC. Fetal tissue dosages of retinoids: experimental study concerning a case of isotretinoin (Roaccutan) administration and pregnancy. Fetal Diagn Ther. 1995;10:189–91.

Biesalski HK. Comparative assessment of the toxicology of vitamin A and retinoids in man. Toxicology. 1989;57:117–61.

Bikowski JB. Mechanisms of the comedolytic and anti-inflammatory properties of topical retinoids. J Drugs Dermatol. 2005;4:41–7.

Bollag W, Ott F. Successful treatment of chronic hand eczema with oral 9-cis-retinoic acid. Dermatology. 1999;199:308–12.

Brewster AM, Lee JJ, Clayman GL, Clifford JL, Reyes MJ, Zhou X, et al. Randomized trial of adjuvant 13-cis-retinoic acid and interferon alfa for patients with aggressive skin squamous cell carcinoma. J Clin Oncol. 2007;25:1974–8.

Chen WC, Sass JO, Seltmann H, Nau H, Orfanos CE, Zouboulis CC. Biological effects and metabolism of 9-cis-retinoic acid and its metabolite 9,13-di-cis-retinoic acid in HaCaT keratinocytes in vitro: comparison with all-trans-retinoic acid. Arch Dermatol Res. 2000;292:612–20.

Chroni E, Monastirli A, Tsambaos D. Neuromuscular adverse effects associated with systemic retinoid dermatotherapy: monitoring and treatment algorithm for clinicians. Drug Saf. 2010; 33:25–34.

Cunliffe WJ, Poncet M, Loesche C, Verschoore M. A comparison of the efficacy and tolerability of adapalene 0.1% gel versus tretinoin 0.025% gel in patients with acne vulgaris: a metaanalysis of five randomized trials. Br J Dermatol. 1998;139 Suppl 52:48–56.

Del Rosso JQ. Study results of benzoyl peroxide 5%/clindamycin 1% topical gel, adapalene 0.1% gel, and use in combination for acne vulgaris. J Drugs Dermatol. 2007;6:616–22.

Downie MM, Sanders DA, Kealey T. Modelling the remission of individual acne lesions in vitro. Br J Dermatol. 2002;147:869–78.

Duvic M. Bexarotene and DAB(389)IL-2 (denileukin diftitox, ONTAK) in treatment of cutaneous T-cell lymphomas: algorithms. Clin Lymphoma. 2000;1 Suppl 1:S51–5.

Duvic M, Hymes K, Heald P, Breneman D, Martin AG, Myskowski P, et al. Bexarotene is effective and safe for treatment of refractory, advanced-stage cutaneous T-cell lymphoma: multinational phase II-m trial results. J Clin Oncol. 2001a;19:2456–71.

Duvic M, Martin AG, Kim Y, et al. Phase 2 and 3 clinical trial of oral bexarotene (Targretin capsules) for the treatment of refractory or persistent early-stage cutaneous T-cell lymphoma. Arch Dermatol. 2001b;137:581–93.

Enders SI, Enders IM. Isotretinoin and psychiatric iIlnessin adolescents and young adults. Ann Pharmacother. 2003;37:1124–7.

Estival JL, Dupin M, Kanitakis J, Combemale P. Capillary leak syndrome induced by acitretin. Br J Dermatol. 2004;150:150–2.

Ethan-Quan, Wolverton. Ligand Pharmaceuticals. Targretin (bexarotene) capsules prescribing information. San Diego, CA. 2003.

Fanjul A, Dawson MI, Hobbs PD, Jong L, Cameron JF, Harlev E, et al. A new class of retinoids with selective inhibition of AP-1 inhibits proliferation. Nature. 1994;372:107–11.

Ganceviciene R, Zouboulis CC. Isotretinoin: state of the art treatment for acne vulgaris. Expert Rev Dermatol. 2007;2:693–706.

Gold JA, Schupack JL, Nemel MA. Ocular side effects of retinoids. Int J Dermatol. 1989;28:218–25.

Goldberg LH, Hsu SH, Alcalay J. Effectiveness of isotretinoin in preventing the appearance of basal cell carcinomas in basal cell nevus syndrome. J Am Acad Dermatol. 1989;21:144–5.

Gollnick H, Cunliffe W, Berson D, Dreno B, Finlay A, Leyden JJ, et al. Management of acne: a report from a Global Alliance to Improve Outcomes in Acne. J Am Acad Dermatol. 2003;49:S1–37.

Gupta AK, Goldfarb MT, Ellis CN, Voorhees JJ. Side-effect profile of acitretin therapy in psoriasis. J Am Acad Dermatol. 1989;20:1088–93.

Guy R, Ridden C, Kealey T. The improved organ maintenance of the human sebaceous gland: modeling in vitro the effects of epidermal growth factor, androgens, estrogens, 13-cis retinoic acid, and phenol red. J Invest Dermatol. 1996;106:454–60.

Hanifi JM, Stevens V, Sheth P, et al. Novel treatment of severe chronic hand dermatitis with bexarotene gel. Br J Dermatol. 2004;150:545–53.

Hassig CA, Fleischer TC, Billin AN, Schreiber SL, Ayer DE. Histone deacetylase activity is required for full transcriptional repression by mSin3A. Cell. 1997;89:341–7.

Haugen BR. Drugs that suppress TSH or cause central hypothyroidism. Best Pract Res Clin Endocrinol Metab. 2009;23:793–800.

Imam A, Hoyos B, Swenson C, Levi E, Chua R, Viriya E, et al. Retinoids as ligands and coactivators of protein kinase C alpha. FASEB J. 2001;15:28–30.

Jones DA. The potential immunomodulatory effects of topical retinoids. Dermatol Online J. 2005;11:3.

Kanekura T, Higashi Y, Kanzaki T. Inhibitory effects of 9-cis-retinoic acid and pyrrolidinedithiocarbamate on cyclooxygenase (COX)-2 expression and cell growth in human skin squamous carcinoma cells. Cancer Lett. 2000;161:177–83.

Karlsson T, Vahlquist A, Kedishvili N, Törmä H. 13-cis-retinoic acid competitively inhibits 3 alpha-hydroxysteroid oxidation by retinol dehydrogenase RoDH-4: a mechanism for its anti-androgenic effects in sebaceous glands? Biochem Biophys Res Commun. 2003;303:273–8.

King K, Jones DH, Daltrey DC, Cunliffe WJ. A double-blind study of the effects of 13-cis-retinoic acid on acne, sebum excretion rate and microbial population. Br J Dermatol. 1982;107: 583–90.

Korting HC, Braun-Falco O. Efficacy and tolerability of combined topical treatment of acne vulgaris with tretinoin and erythromycin in general practice. Drugs Exp Clin Res. 1989;15:447–51.

Kovach BT, Murphy G, Otley CC, Shumack S, Ulrich C, Stasko T. Oral retinoids for chemoprevention of skin cancers in organ transplant recipients: results of a survey. Transplant Proc. 2006;38:1366–8.

Kraemer KH, DiGiovanna JJ, Moshell AN, Tarone RE, Peck GL. Prevention of skin cancer in xeroderma pigmentosum with the use of oral isotretinoin. N Engl J Med. 1988;318:1633–7.

Lavker RM, Leyden JJ, Thorne EG. An ultrastructural study of the effects of topical tretinoin on microcomedones. Clin Ther. 1992;14:773–80.

Layton AM, Dreno B, Gollnick HP, Zouboulis CC. A review of the European Directive for prescribing systemic isotretinoin for acne vulgaris. J Eur Acad Dermatol Venereol. 2006;20:773–6.

Lippman SM, Kesseier JF, Meyskens FL Jr. Retinoids as preventive and therapeutic anticancer agents (Part 11). Cancer Treat Rep. 1987;21:493–515.

Marazzi P, Boorman GC, Donald AE, Davies HD. Clinical evaluation of double strength isotrexin versus benzamycin in the topical treatment of mild to moderate acne vulgaris. J Dermatol Treat. 2002;13:111–7.

Mills OH Jr, Kligman AM. Treatment of acne vulgaris with topically applied erythromycin and tretinoin. Acta Derm Venereol. 1978;58:555–7.

Nelson AM, Gilliland KL, Cong Z, Thiboutot DM. 13-cis Retinoic acid induces apoptosis and cell cycle arrest in human SEB-1 sebocytes. J Invest Dermatol. 2006;126:2178–89.

Nguyen EH, Wolverton SE. Systemic retinoids. In: Wolverton SE, Wilkin JK, editors. Systemic drugs for skin diseases. Philadelphia: WB Saunders; 2000. p. 269–310.

Olsen EA, Weed WW, Meyer CJ, Cobo LM. A double-blind, placebo-controlled trial of acitretin for the treatment of psoriasis. J Am Acad Dermatol. 1989;21:681–6.

Orfanos CE, Zouboulis CC. Oral retinoids in the treatment of seborrhoea and acne. Dermatology. 1998;196:140–7.

Orfanos CE, Zouboulis CC, Almond-Roesler B, Geilen C. Current use and future potential role of retinoids in dermatology. Drugs. 1997;53:358–88.

Otley CC, Stasko T, Tope WD, Lebwohl M. Chemoprevention of nonmelanoma skin cancer with systemic retinoids: practical dosing and management of adverse effects. Dermatol Surg. 2006;32:562–8.

Papakonstantinou E, Aletras AJ, Glass E, Tsogas P, Dionyssopoulos A, Adjaye J, et al. Matrix metalloproteinases of epithelial origin in facial sebum of patients with acne and their regulation by isotretinoin. J Invest Dermatol. 2005;125:673–84.

Rollman O, Vahlquist A. Oral isotretinoin (13-cis-retinoic acid) therapy in severe acne: drug and vitamin A concentrations in serum and skin. J Invest Dermatol. 1986;86:384–9.

Rook AH, Jaworsky C, Nguyen T, Grossman RA, Wolfe JT, Witmer WK, et al. Beneficial effect of low-dose systemic retinoid in combination with topical tretinoin for the treatment and prophylaxis of premalignant and malignant skin lesions in renal transplant recipients. Transplantation. 1995;59:714–9.

Ruzicka T, Larsen FG, Galewicz D, Horváth A, Coenraads PJ, Thestrup-Pedersen K, et al. Oral alitretinoin (9-cis-retinoic acid) therapy for chronic hand dermatitis in patients refractory to standard therapy: results of a randomized, double-blind, placebo-controlled, multicenter trial. Arch Dermatol. 2004;140:1453–9.

Scheinfeld N, Bangalore S. Facial edema induced by isotretinoin use: a case and a review of the side effects of isotretinoin. J Drugs Dermatol. 2006;5:467–8.

Shuttleworth D, Marks R, Griftin PJ, Salaman JR. Treatment of cutaneous neoplasia with etretinate in renal transplant recipients. Q J Med. 1988;68:717–25.

Smit JW, Stokkel MP, Pereira AM, Romijn JA, Visser TJ. Bexarotene-induced hypothyroidism: bexarotene stimulates the peripheral metabolism of thyroid hormones. J Clin Endocrinol Metab. 2007;92:2496–9.

Stephensen CB, Rasooly R, Jiang X, Ceddia MA, Weaver CT, Chandraratna RA, et al. Vitamin A enhances in vitro Th2 development via retinoid X receptor pathway. J Immunol. 2002;168:4495–503.

Sun SY, Yue P, Mao L, Dawson MI, Shroot B, Lamph WW, et al. Identification of receptor-selective retinoids that are potent inhibitors of the growth of human head and neck squamous cell carcinoma cells. Clin Cancer Res. 2000;6:1563–73.

Tanghetti E, Abramovits W, Solomon B, Loven K, Shalita A. Tazarotene versus tazarotene plus clindamycin/ benzoyl peroxide in the treatment of acne vulgaris: a multicenter, doubleblind, randomized parallel-group trial. J Drugs Dermatol. 2006;5:256–61.

Thiboutot DM, Shalita AR, Yamauchi PS, Dawson C, Arsonnaud S, Kang S. Combination therapy with adapalene gel 0.1% and doxycycline for severe acne vulgaris: a multicenter, investigator-blind, randomized, controlled study. Skinmed. 2005;4:138–46.

Thiboutot DM, Weiss J, Bucko A, Eichenfield L, Jones T, Clark S, et al. Adapalene-benzoyl peroxide, a fixed-dose combination for the treatment of acne vulgaris: results of a multicenter, randomized double-blind, controlled study. J Am Acad Dermatol. 2007;57:791–9.

Thielitz A, Sidou F, Gollnick H. Control of microcomedone formation throughout a maintenance treatment with adapalene gel, 0.1%. J Eur Acad Dermatol Venereol. 2007;21:747–53.

Thomas X, Anglaret B, Thiebaut A, Belhabri A, Treille-Ritouet D, Fiere D, Archimbaud E. Improvement of prognosis in refractory and relapsed acute promyelocytic leukemia over recent years: the role of all-trans retinoic acid therapy. Ann Hematol. 1997;75:195–200.

Tsukada M, Schröder M, Roos TC, Chandraratna RA, Reichert U, Merk HF, Orfanos CE, Zouboulis CC. 13-cis retinoic acid exerts its specific activity on human sebocytes through selective intracellular isomerization to all-trans retinoic acid and binding to retinoid acid receptors. J Invest Dermatol. 2000;115:321–7.

Tsukada M, Schröder M, Seltmann H, Orfanos CE, Zouboulis CC. High albumin levels restrict the kinetics of 13-cis retinoic acid uptake and intracellular isomerization to all-trans retinoic acid and inhibit its anti-proliferative effect on SZ95 sebocytes. J Invest Dermatol. 2002;119: 182–5.

Vane FM, Bugge CJL, Rodriguez LC, Rosenberger M, Doran TI. Human biliary metabolites of isotretinoin: identification, quantification, synthesis, and biological activity. Xenobiotica. 1990;20:193–207.

Walmsley S, Northfelt DW, Melosky B, Conant M, Friedman-Kien AE, Wagner B. Treatment of AIDS related cutaneous Kaposi's sarcoma with topical alitretinoin (9-cis-retinoic acid) gel. Panretin Gel North American Study Group. J Acquir Immune Defic Syndr. 1999;22:235–46.

Webster GF, Guenther L, Poulin YP, Solomon BA, Loven K, Lee J. A multicenter, double-blind, randomized comparison study of the efficacy and tolerability of once-daily tazarotene 0.1% gel and adapalene 0.1% gel for the treatment of facial acne vulgaris. Cutis. 2002;69:4–11.

Wiegand UW, Chou RC. Pharmacokinetics of acitretin and etretinate. J Am Acad Dermatol. 1998;39:S25–33.

Zane LT, Leyden WA, Marqueling AL, Manos MM. A population-based analysis of laboratory abnormalities during isotretinoin therapy for acne vulgaris. Arch Dermatol. 2006;142:1016–22.

Zhang JZ, Li LF, Tu YT, Zheng J. A successful maintenance approach in inflammatory acne with adapalene gel 0.1% after an initial treatment in combination with clindamycin topical solution 1% or after monotherapy with clindamycin topical solution 1%. J Dermatol Treat. 2004;15:372–8.

Zouboulis CC, Korge BP, Mischke D, Orfanos CE. Altered proliferation, synthetic activity, and differentiation of cultured human sebocytes in the absence of vitamin A and their modulation by synthetic retinoids. J Invest Dermatol. 1993;101:628–33.

Zouboulis CC, Orfanos CE. Retinoids. In: Millikan LE, editor. Drug therapy in dermatology. New York/Basel: Marcel Dekker; 2000. p. 171–233.

Zouboulis CC. Retinoids – which dermatological indications will benefit in the near future? Skin Pharmacol Appl Skin Physiol. 2001;14:303–15.

Zouboulis CC. Isotretinoin revisited: pluripotent effects on human sebaceous gland cells. J Invest Dermatol. 2006a;126:2154–6.

Zouboulis CC. The truth behind this undeniable efficacy – recurrence rates and relapse risk factors of acne treatment with oral isotretinoin. Dermatology. 2006b;212:99–100.

Zouboulis CC, Rabe T. Hormonal antiandrogens in acne treatment. J Dtsch Dermatol Ges. 2010;8 Suppl 1:S60–74.

Chapter 3
Relevance of the Cutaneous Vitamin D Endocrine System for Skin Physiology and Treatment of Skin Diseases

Léa Trémezaygues and Jörg Reichrath

Core Messages

- Vitamin D can be absorbed from the diet or synthesized in the skin from 7-dehydrocholesterol under the influence of ultraviolet B radiation.
- The keratinocyte is the only cell type known today that is able to synthesize the biologically active vitamin D metabolite 1,25$(OH)_2D_3$ from 7-dehydrocholesterol.
- Numerous in vitro and in vivo studies have demonstrated dose-dependent effects of vitamin D analogues on cell proliferation and differentiation as well as immunomodulatory effects, effects on apoptosis, and antioxidative and cyto-protective effects.
- 1,25$(OH)_2D_3$ and numerous of its analogues are used in the treatment of psoriasis and other skin diseases.

3.1 Introduction

Our understanding of the relevance of the vitamin D metabolism in human skin has significantly improved during the last few years. The skin is the only organ that is capable of synthesizing vitamin D after exposure to sunlight. On the other hand, ultraviolet (UV) exposure is one of the major risk factors for epithelial skin cancer. Thus, there is a conflict between the positive and negative properties of sunlight. Today it is commonly accepted that vitamin D deficiency is not only associated with affecting bone and calcium metabolism but also with an elevated risk for multiple other diseases (e.g., various cancers, cardiovascular diseases, infections,

L. Trémezaygues (✉)
Department of Dermatology, The Saarland University Hospital,
Kirrberger Strasse, 66421 Homburg/Saar, Germany
e-mail: maillea1904@aol.com

A. Pappas (ed.), *Nutrition and Skin: Lessons for Anti-Aging, Beauty and Healthy Skin*,
DOI 10.1007/978-1-4419-7967-4_3,

autoimmune diseases). In this chapter we summarize current scientific knowledge about the importance of vitamin D metabolism in skin. Furthermore, we elucidate new aspects of the use of vitamin D and its analogue in dermatology, especially in regard to preventing skin cancer and treating psoriasis.

3.2 Vitamin D Synthesis in Skin

Vitamin D, the precursor of the biologically active vitamin D metabolite 1,25-dihydroxyvitamin D [1,25(OH)$_2$D] can be absorbed from the diet or synthesized in skin under the influence of UVB radiation from 7-dehydrocholesterol (7-DHC). All in all, nine enzymatic reactions are involved in the photochemical cutaneous synthesis of vitamin D, hereunder four photoreversible reactions and one nonreversible phototransformation (Lehmann et al. 2001). Whereas vitamin D$_2$ (ergocalciferol) can be found in plants, vitamin D$_3$ (cholecalciferol) is photochemically synthesized under the influence of UVB radiation in the skin of animals and humans.

The biologically active vitamin D metabolite 1,25(OH)$_2$D, or calcitriol, which circulates in the blood, is synthesized from vitamin D by a well-characterized biochemical reaction cascade. First, it is hydroxylated in the liver in the C-25 position by a cytochrome P450 enzyme, the vitamin D-25-hydroxylase (CYP27A1), before it is hydroxylated a second time in the kidney in the C-1 position by another cytochrome P450 enzyme, 25-hydroxyvitamin D-1α-hydroxylase (CYP27B1) (Holick 2007).

Production of 1,25-(OH)$_2$D$_3$ in the kidney is regulated by a feedback mechanism of the hormone itself as well as by parathyroid hormone, calcium, and cytokines such as interferon-γ (IFNγ) and tumor necrosis factor-α (TNFα) (Holick 2007). During the 1970s, it was concluded that the kidney was the sole source of 1,25(OH)$_2$D$_3$ production. However, more recent in vitro trials and studies of anephric humans demonstrated that cultured human keratinocytes, monocytes, macrophages, osteoblasts, and prostate and colon cells, among others, express enzymatic machinery for the synthesis of 1,25(OH)$_2$D$_3$ (i.e., 5-hydroxyvitamin D-1α-hydroxylase) and are thus able to synthesize 1,25-(OH)$_2$D$_3$ (Bikle et al. 1986; Holick 2003, 2007). In keratinocytes, studies could even prove the presence of 1α-hydroxylase (CYP27A1) and 25-hydroxylase (CYP27B1) (Fu et al. 1997; Lehmann et al. 1999). According to these findings, the keratinocyte was the only cell type known today that is able to synthesize 1,25(OH)$_2$D$_3$ from 7-DHC.

The metabolism of 1,25(OH)$_2$D circulating in the blood to calcitroic acid results from another well-characterized biochemical reaction cascade. The reduction of 1,25(OH)$_2$D$_3$ to 24,25-dihydroxycholecalciferol in the kidney and consecutive reactions are effectuated by a third cytochrome P450 enzyme, 1,25-dihydroxyvitamin D-24-hydroxylase (CYP24A1) (Holick 2007). 24,25-Dihydroxycholecalciferol has only slight biological activity. According to our present knowledge 1,25(OH)$_2$D$_3$ has 100- to 1,000-fold higher biological activity than other natural vitamin D metabolites (Bikle et al. 1994).

3.3 Biological Effects of 1α,25-$(OH)_2D_3$ in the Skin

3.3.1 Inhibition of Proliferation and Induction of Differentiation in Keratinocytes

Numerous in vitro and in vivo studies have demonstrated dose-dependent effects of vitamin D analogues on cell proliferation and differentiation. At low concentrations ($<10^{-8}$ M), calcitriol promotes proliferation of keratinocytes in vitro; at higher pharmacological doses ($\geq 10^{-8}$ M) keratinocyte proliferation is inhibited (Gniadecki 1996). In psoriatic skin, immunohistochemical and biochemical analysis have clearly shown antiproliferative and differentiation-inducing effects in epidermal keratinocytes along with treatment with vitamin D analogues in vivo (Reichrath et al. 1997a,b). It has been demonstrated that the immunohistochemical staining pattern for various markers of epidermal proliferation [e.g., proliferating cell nuclear antigen (PCNA), Ki-67-antigen] and differentiation (e.g., involucrin, transglutaminase K, filaggrin, cytokeratin 10) changes in lesional psoriatic skin along with topical treatment with vitamin D analogues almost completely to the staining pattern characteristic for nonlesional psoriatic or normal skin (Reichrath et al. 1997a,b).

Although the mechanisms that underlie the antiproliferative and differentiation-inducing effects of vitamin D analogues on keratinocytes are not completely understood, it is well known that these effects are at least in part genomic and mediated via vitamin D receptors (VDRs). It has been shown that keratinocytes from VDR-deficient mice do not respond in vitro to the antiproliferative effects of vitamin D analogues. In lesional psoriatic skin, the clinical improvement correlates with an increase of VDR mRNA in calcitriol-treated skin areas (Holick and Reichrath 1999). However, not all patients with psoriasis respond to treatment with vitamin D analogues. It has been demonstrated that a "responder" can be discriminated from a "nonresponder" by an increase in VDR mRNA in treated skin areas (Holick and Reichrath 1999). These data suggest that the ability of calcitriol to regulate keratinocyte growth is closely linked to the expression of VDRs. The target genes of topical calcitriol that are responsible for its therapeutic efficacy in psoriasis are still unknown. Major candidates for calcitriol target genes that are responsible for the calcitriol-induced terminal differentiation in keratinocytes are distinct cell cycle-associated proteins (i.e., INK4 family), including p21/WAF-1 (Holick and Reichrath 1999).

3.3.2 Immunmodulatory Effects in the Skin

During the last few years, important new immunomodulatory effects of vitamin D analogues have been characterized (Weber et al. 2005; Holick 2003, 2007). It has been demonstrated that various cell types involved in immunological reactions (e.g., monocytes, T and B lymphocytes, Langerhans cells) not only express VDR, they possess the enzymatic machinery (25-hydroxyvitamin D-1α-hydroxylase) for local synthesis of calcitriol (Holick 2007). Today, local synthesis of calcitriol in immune cells is considered to be of high importance for the regulation and control of various immune responses. 1α,25-$(OH)_2D$ inhibits activation of T cells and induces the generation of CD25+/CD4+ regulatory T cells (Holick 2007). In dendritic cells, calcitriol inhibits maturation and induces a phenotype that promotes tolerance and inhibits immunity after stimulation with antigen (Holick 2003, 2007). In dendritic cells, calcitriol suppresses expression of major histocompatibility complex II (MHC II) molecules and of co-stimulatory molecules, including CD40, CD80, and CD86 (Griffin and Kumar 2003). In these cells, production of interleukin −10 (IL-10) is stimulated and production of IL-12 inhibited, leading to suppression of T-cell activation (Griffin and Kumar 2003).

Impressive therapeutic effects were seen after application of vitamin D analogues for diseases that are related with the function of T cells or dendritic cells (experimentally induced allergic encephalomyelitis, collagen-induced arthritis, autoimmune thyreoiditis, diabetes mellitus type I, graft-versus-host reaction) (Holick 2007). Other studies have shown that vitamin D deficiency encourages the pathogenesis of many autoimmune diseases (e.g., diabetes mellitus type I) and that a sufficient vitamin D serum concentration reduces the incidence of those diseases. At present, a connection between vitamin D and pathogenesis of atopic dermatitis is being discussed. Epidemiological studies have indicated that patients with atopic dermatitis have a lower vitamin D intake than their controls (Solvoll et al. 2000). It has been demonstrated that vitamin D analogues suppress in vitro immunoglobulin E (IgE) production and IgE-mediated cutaneous reactions (Katayama et al. 1996). These immunomodulatory effects identify vitamin D analogues—most likely new vitamin D analogues with selective immunomodulatory activity—as promising new drugs for the prevention and treatment of inflammatory skin diseases including atopic dermatitis and allergic contact dermatitis.

New insights have demonstrated that calcitriol induces the expression of the CCR-10 receptor on the surface of T cells, which leads to the migration of the T cells toward CCL-27-expressing epidermal keratinocytes (Sigmundsdottir et al. 2007). This UVB-induced and vitamin D-mediated T-cell mobilization from the blood vessels of the dermis into the epidermis characterizes another immunomodulatory effect of vitamin D: an on-demand increase of the T-cell answer in the epidermis (Sigmundrdottir et al. 2007). The clinical relevance of this function of vitamin D has not yet been totally clarified and remains the subject of further studies.

3.3.3 Regulation of Apoptosis

1,25-Dihydroxyvitamin D_3 has the potential to induce the neutral Mg^{2+}-dependent sphingomyelinase, which hydrolyzes sphingomyelin to ceramide (Okasaki et al. 1989). Interestingly, ceramide stimulates the prodifferentiating effect from

1,25-dihydroxyvitamin D_3 on keratinocytes (Bielawska et al. 1992). Moreover, it plays an important role in the induction of apoptosis of a variety of cells, including keratinocytes (Geilen et al. 1997). It has been demonstrated that physiological concentrations of 1,25-dihydroxyvitamin D_3 in keratinocyte cultures do not induce apoptosis. To the contrary, they generate apoptosis resistance against ceramides, UV radiation, and TNFα (Manggau et al. 2001). The cytoprotective/antiapoptotic effect of 1,25-dihydroxyvitamin D_3 is obviously linked to the development of sphingosine-1-phosphate. This is also clarified by the fact that the antiapoptotic effect of 1,25-dihydroxyvitamin D_3 can be completely suppressed by addition of the sphingosine kinase inhibitor *N,N*-dimethylsphingosine (Manggau et al. 2001). In contrast, pharmacological concentrations of 1,25-dihydroxyvitamin D_3 ($\geq 10^{-6}$ M) do induce apoptosis. Similar effects have been observed in the regulation of the keratinocyte growth: As outlined above, concentrations of 1,25-dihydroxyvitamin D_3 around 10^{-11} M stimulate cell proliferation, whereas higher concentrations have a dose-dependent antiproliferative effect (Gniadecki 1996).

3.3.4 Cytoprotective Effects

The development of many types of skin cancer, including basal cell carcinoma (BCC) and squamous cell carcinoma (SCC), is induced by UV exposure (Elwood and Jopson 1997). Whereas the total UV dose accumulated in the exposed skin area during a lifetime is responsible for the genesis of SCCs (Alam and Ratner 2001), the origin of BCCs is presumed to be both in the accumulated UV dose in the exposed skin area over a lifetime and in higher intermittent UV doses (Kricker et al. 1994). In contrast, the development of malignant melanoma may be the result of higher intermittent UV exposure during early childhood (0–6 years) (Osterlind et al. 1988). Counted among these exposures are already scattered suberythemal solar exposures and, above all, mild and heavy sunburns. They lead to the appearance of multiple additional nevi, which are considered a risk factor for the development of malignant melanoma in adults (Osterlind et al. 1988). Excessive UV exposure, especially UVB with a wavelength range of 290–320 mn, is consequently a major risk factor in the development of skin cancer. On the other hand, sufficient doses of UVB are needed for the photochemical synthesis of vitamin D in the skin (Reichrath 2006). This observation led to the hypothesis that cutaneous vitamin D synthesis may act as an evolutionary highly conserved natural protection system against the hazardous effects of UVB radiation in the skin (Trémezaygues 2009).

Many studies have confirmed that 1,25-dihydroxyvitamin D_3 protects human keratinocytes at least partly against UVB-induced cell damage (Lee and Youn 1998; Gupta et al. 2006; De Haes et al. 2004; Trémezaygues 2009). The underlying mechanisms are manifold and are mainly associated with suppression of UVB-induced apoptosis—reduction of cytochrome c-releasing, decreased poly(adenosine diphosphate ribose) polymerase (PARP) cleavage (Lee and Youn 1998). The research group of De Haes et al. verified the cytoprotective effects of 1,25-dihydroxyvitamin

D_3 in UVB-irradiated keratinocytes using morphological and colorimetric vitality assays (De Haes et al. 2004, 2005). A significant cytoprotective effect resulted, starting from a 1,25-dihydroxyvitamin D_3 concentration of 10^{-8} M and pretreatment with 1,25-dihydroxyvitamin D_3 over at least 8 h. 1,25-Dihydroxyvitamin D_3 doses of 10^{-6}–10^{-8} M were necessary to obtain a considerable cytoprotective effect (De Haes et al. 2004, 2005). Using an enzyme-linked immunosorbent assay (ELISA) that detects DNA fragmentation, they showed that pretreatment of the keratinocytes with 1,25-dihydroxyvitamin D_3 (1 μM) over 24 h suppresses UVB-induced apoptosis up to 55–70% (De Haes et al. 2004, 2005). They also showed that pretreatment of keratinocytes with 1,25-dihydroxyvitamin D_3 (1 μM) reduces the mitochondrial cytochrome c release—a marker of UVB-induced apoptosis—up to 90%. Two important mediators of the UV answer—activation of the -Jun-NH_2-terminal kinase and production of IL-6—are also reduced about 30% (activation of the -Jun-NH_2-terminal kinase) respective 75–90% (production of IL-6) by a pretreatment with 1,25-dihydroxyvitamin D_3 (1 μM) (De Haes et al. 2004, 2005). Even the UVB-induced cleavage of PARP is inhibited by 24 h of pretreatment of the cells with 1,25-dihydroxyvitamin D_3 (1 μM) (De Haes et al. 2004, 2005). Furthermore, it could be demonstrated that pretreatment of the keratinocytes with 1,25-dihydroxyvitamin D_3 leads to induction of metallothionein (MT) mRNA. MT is an antioxidant and acts as a radical catcher after UV irradiation (Lee and Youn 1998; De Haes et al. 2004, 2005). Moreover, 1,25-dihydroxyvitamin D_3 protects keratinocytes by induction of antiapoptotic proteins such as Bcl-2 and activation of the MEK/ERK and PI-3 K/Akt metabolic pathways before apoptosis (De Haes et al. 2005). Experiments of Trémezaygues et al., using a colony-forming-unit culture assay, proved that after 48 h of pretreatment of keratinocytes with 1,25-dihydroxyvitamin D_3 (10^{-7} M) the number of cell colonies counted after a single irradiation with 100 J/cm^2 and a growth period of 7 days is twice higher as that of controls that were not treated with vitamin D (Trémezaygues 2009). Furthermore, it was demonstrated that 1,25-dihydroxyvitamin D_3 (10^{-7} M) has a photoprotective effect in WST-1- and crystal violet-based assays after irradiation of the keratinocytes with ascending doses of UVB light (100–1,000 J/cm^2). Indeed, ascending UVB doses lead to a decrease of the NADH levels (WST-1 assay) in respect to a decrease of crystal violet absorption (crystal violet assay) in the keratinocytes, whereas the calcitriol-treated keratinocytes show higher NADH levels (WST-1 assay) and higher crystal violet absorption than the controls. The NADH level and crystal violet absorption are measures of the metabolic activity and the vitality of the cells. Thus, these experiments confirmed a cytoprotective effect of 1,25-dihydroxyvitamin D_3 against the hazardous effects of UVB light (Trémezaygues 2009).

It is well recognized that the photocarcinogenesis of skin cancer is mainly due to mutations resulting from insufficiently repaired DNA photoproducts (De Haes et al. 2004). The most established DNA photoproducts caused by UV irradiation are cyclobutane pyrimidine dimers (CPDs). New research shows that 1,25-dihydroxyvitamin D_3 prevents human keratinocytes from the induction of CPDs after UVB irradiation (De Haes et al. 2004; Trémezaygues 2009). Gupta et al. described a reduction of the number of CPDs after a 24-h pretreatment with 1,25-dihydroxyvitamin D_3 (10^{-9} M) followed by irradiation of the cells with UVB 200 mJ/cm^2 compared to that of the controls, which were not treated with vitamin D (Gupta et al. 2006). In line with these results, Trémezaygues demonstrated that 48-h pretreatment of keratinocytes with 1,25-dihydroxyvitamin D_3 (10^{-7} M) has a protective effect on the cells, even after irradiation with higher doses of UVB (100 and 1,000 J/cm^2) (Trémezaygues 2009). Complementing the results of Gupta et al., Trémezaygues showed that after pretreatment of the cells with 1,25-dihydroxyvitamin D_3, in addition to a reduced number of formed CPDs, there was quicker reparation of the CPDs in 1,25-dihydroxyvitamin D_3-pretreated HaCaT keratinocytes than in the controls, which were not treated with vitamin D (Trémezaygues 2009).

Regarding the influence of vitamin D metabolites on the development of radiation damage, a number of investigations were performed during the last few years. The characteristic damage after ionizing radiation is the double-strand break. The histone protein H2AX is phosphorylated in position 139 at the carboxy terminus as an answer to a double-strand break; the result is γH2AX. This phosphorylated histone protein recruits retains various repair proteins at the site of the double strand break. It can therefore be considered as a marker for double-strand breaks. Recent studies have shown reduced immunoreactivity for γH2AX caused by ionizing radiation after pretreating the cells with 1,25-dihydroxyvitamin D_3. A significant cytoprotective effect was proved after 48 h of pretreatment of the cells with 1,25-dihydroxyvitamin D_3 at a concentration of 10^{-7} M (Trémezaygues 2009).

To summarize, the current literature speaks in favor of a cytoprotective effect of vitamin D, for which the clinical potential does not yet seem to be exhausted.

3.3.5 *Antioxidative Effects*

1,25-Dihydroxyvitamin D_3 has a photoprotective effect on keratinocytes in vitro. It induces production of the protein metallothionein in keratinocytes, whose antioxidative potential has been described (Lee and Youn 1998; Hanada et al. 1995). This may represent an important mechanism, protecting the cells against the UVB-induced synthesis of reactive oxygen radicals.

3.4 Clinical Studies of Vitamin D and Its Analogues in Psoriasis and Other Skin Diseases

Because there is no cure for psoriasis at present, therapeutic strategies are intended to achieve and maintain remission—i.e., reduction of the extent (percentage of the body area involved) and severity (degree of erythema, scaling, plaque elevation)—of the disease by minimizing adverse events. The importance of $1,25(OH)_2D_3$ and its analogues for the treatment of psoriasis resulted from two independent lines of investigation. Because psoriasis is a hyperproliferative skin disorder, it seemed rational that the antiproliferative effects of calcitriol could be used for the treatment of this disease. Before launching clinical trials in 1985, MacLaughlin and associates reported that psoriatic fibroblasts were partially resistant to the antiproliferative effects of $1,25(OH)_2D_3$ (MacLaughlin et al. 1985). This observation led these researchers to speculate that calcitriol may be effective in the treatment of hyperproliferative skin diseases such as psoriasis. The second line of investigation resulted from a clinical observation. In 1985, Morimoto and Kumahara reported that a patient who was treated orally with 1α-hydroxyvitamin D_3 for osteoporosis had a drastic remission of psoriatic skin lesions (Morimoto and Kumahara 1985). In a follow-up study, they demonstrated that almost 80% of 17 patients with psoriasis who were treated orally with 1α-hydroxyvitamin D_3 at a dose of 1.0 μg/day for up to 6 month showed clinically significant improvement (Morimoto et al. 1986).

Up to now, various studies reported that numerous vitamin D analogues, including calcitriol, calcipotriol, tacalcitol, maxacalcitol, and beocalcidiol, are effective and safe for topical treatment of psoriasis (Holick et al. 1996; Perez et al. 1996a; Kragballe et al. 1988; van de Kerkhof et al. 1989; Helfrich et al. 2007). It was demonstrated that topical calcitriol is highly effective and safe for long-term treatment of psoriasis vulgaris (Perez et al. 1996a). In clinical trials, calcipotriol ointment and cream reduced the mean Psoriasis Area and Severity Index (PASI) scores by 55–72% and 49–50%, respectively, after treatment for 6–8 weeks (Van de Kerkhof and Vissers 2003). Applied topically twice a day in amounts of up to 100 g ointment (50 μg calcipotriol/g ointment) per week, calcipotriol, the synthetic analogue of calcitriol, was shown to be more effective for topical treatment of psoriasis than betamethasone 17-valerate ointment (Kragballe et al. 1991). Topical calcipotriol has been compared with other topical treatments for psoriasis. Calcipotriol cream has been reported to be as effective and cosmetically more favorable than coal tar in a small, observer-blinded trial in patients with psoriasis (Tzaneva et al. 2003).

Recently, twice-daily calcipotriol ointment was compared with once-daily short-contact dithranol cream therapy in a randomized controlled trial (RCT) of supervised treatment of psoriasis in a daycare setting (De Korte et al. 2008). This multicenter RCT, which was performed in six centers in The Netherlands, included 106 patients with chronic plaque psoriasis. Among them, 54 received calcipotriol ointment twice daily, and 52 were given dithranol cream once daily during a 12-week intensive treatment program. Patients were treated at the daycare center using the care instruction principle of daily visits during the first week and twice-weekly visits subsequently for up to 12 weeks. Quality of life (QOL) was assessed with the Skindex-29 and the Medical Outcomes Study 36-item Short-Form General Health Survey (SF-36). No statistically significant differences were found between the calcipotriol and the dithranol group in any of the QOL domains or scales of the Skindex-29 and the SF-36 at the end of treatment.

Calcipotriol is available in three formulations: cream, ointment, and solution. In a large, randomized, double-blind, controlled trial, twice-daily application of calcipotriol cream was significantly more effective than once-daily application of calcipotriol cream in terms of the mean percentage reduction in PASI from baseline (48.3% vs. 40.6%, $P = 0.006$) (Kragballe et al. 1998). In this study, the reduction in PASI with twice-daily application of calcipotriol cream did not differ from that with application of calcipotriol cream in the morning plus clobetasone butyrate cream in the evening (53.7%) and was significantly lower than that with application of calcipotriol cream in the morning plus betamethasone valerate cream in the evening (57.5%).

Lately, an investigator-masked, randomized, multicenter comparison of the efficacy and safety of twice-daily applications of calcitriol 3 μg/g ointment versus calcipotriol 50 μg/g ointment in patients with mild to moderate chronic plaque-type psoriasis has been reported (Zhu et al. 2007). In this study, a total of 250 patients of both sexes with mild to moderate chronic plaque-type psoriasis received either calcitriol or calcipotriol ointment twice a day for 12 weeks. Efficacy evaluations comprehend global improvement (on a four-point scale: 0 indicating no change or worse to 3 indicating clear or almost clear) assessed by the investigator and by the subject. Efficacy further included the "dermatological sum score" at each study visit. Safety evaluations included adverse event reporting, cutaneous safety assessed by the investigator, and cutaneous discomfort assessment by the patient (both were on a five-point scale from 0 indicating no discomfort to 4 indicating severe discomfort). At week 12, the LSMean score of global improvement measured by the investigator was 2.27 for calcitriol and 2.22 for calcipotriol. This difference was not statistically significant, demonstrating that calcitriol was not inferior to calcipotriol in terms of global improvement. This same parameter was also scored by the patient, with a mean of 2.12 for calcitriol and 2.09 for calcipotriol. The percentage of patients with at least marked improvement tended to be in favor of calcitriol (95.7% vs. 85.0% for calcipotriol), but the differences were not statistically significant. The mean worst score for the cutaneous safety assessment was higher in the calcipotriol group (0.3 vs. 0.1 and 0.4 vs. 0.2, by the investigator and the patient, respectively). These differences were statistically significant in favor of a superior safety profile for calcitriol ($P=0.0035$). A total of 14 dermatological and treatment-related adverse events were reported with calcipotriol whereas only five were reported with calcitriol for a total of 22 adverse events reported throughout the study. The authors concluded that calcitriol applied twice daily over a 12-week treatment period demonstrated efficacy similar to that of calcipotriol, while showing a significantly better safety profile. It has been noted that a mild dermatitis is seen in about 10% of patients treated with calcipotriol (50 μg/g), particularly on the face (Serup 1994). This side effect was not reported after topical treatment with calcitriol.

More recently, a randomized, placebo-controlled, double-blind, multicentric study analyzing the efficacy and safety of topical becocalcidiol for the treatment of psoriasis vulgaris has been reported (Helfrich et al. 2007). Becocalcidiol is a vitamin D analogue that did not cause hypercalcemia or significant irritation in preclinical trials. In that study, the efficacy and safety of two dosing regimens of becocalcidiol ointment (low dose 75 μg/g once a day for 8 weeks; high dose 75 μg/g twice a day for 8 weeks) in the treatment of plaque-type psoriasis were evaluated. In all, 185 patients with chronic plaque-type psoriasis affecting 2–10% of their body surface area took part in a multicentric, double-blind, parallel-group, vehicle-controlled RCT comparing topical application of placebo, becocalcidiol 75 μg/g once daily (low dose) or becocalcidiol twice daily (high dose) for 8 weeks. Main outcomes included the Physician's Static Global Assessment of Overall Lesion

Severity (PGA) score, the Psoriasis Symptom Severity (PSS) score, adverse events, and laboratory assessment. In this study, in the intention-to-treat population at week 8, high-dose becocalcidiol was statistically superior to the vehicle [$P = 0.002$; 95% confidence interval (CI) 6.7–32.2], with 16 of 61 (26%) patients achieving a PGA score of clear or almost clear. Greater improvement in the PSS score was seen with high-dose becocalcidiol than with the vehicle, but this result did not quite achieve statistical significance ($P = 0.052$; 95% CI −16.2–0.1). In all groups, therapy was well tolerated and safe, with fewer patients experiencing irritation than is reported in studies using calcipotriol. The authors concluded that treatment with high-dose topical becocalcidiol for 8 weeks led to almost or complete clearing of moderate plaque-type psoriasis in more than one-fourth of the patients and that the therapy was safe and well tolerated (Helfrich et al. 2007).

In 1996, a long-term follow-up study reported the efficacy and safety of oral calcitriol for treating psoriasis (Perez et al. 1996b). Among the 85 patients included in the study who received oral calcitriol, 88.0% had some alleviation of their disease after 36 months; 26.5%, 26.3%, and 25.3% had complete, moderate, and slight improvement, respectively. Serum calcium concentrations and 24-h urinary calcium excretion increased by 3.9% and 148.2%, respectively, but were not outside the normal range. Bone mineral density of these patients remained constant.

An important point when considering the use of orally administered calcitriol is the dosing technique. To avoid its effects on enhancing dietary calcium absorption, it is extremely important to provide calcitriol at night time. Perez et al. (1996b) showed that as a result of this dosing technique doses of 2–4 μg/night are well tolerated by psoriatic patients. Lately, the combination of acitretin and oral calcitriol for successful treatment of plaque-type psoriasis has been reported (Ezquerra et al. 2007).

Patients with psoriasis may need intermittent treatment for their entire lives. Vitamin D analogues have been demonstrated not to cause tachyphylaxis during treatment of psoriatic lesions and can therefore be used indefinitely. They are effective and safe for the treatment of skin areas that are usually difficult to treat in psoriatic patients and that respond slowly. Furthermore, vitamin D analogues are effective in the treatment of psoriatic skin lesions in children and in human immunodeficiency virus (HIV)-infected patients.

3.5 Treatment

3.5.1 Scalp Psoriasis

It was shown in a double-blind, randomized multicentric study that calcipotriol solution is an effective topical treatment for scalp psoriasis (Green et al. 1994). A total of 49 patients were treated twice a day over a 4-week period. In all, 60% of the patients treated with calcipotriol showed clearance or marked improvement of their scalp psoriasis versus 17% in the placebo group. No side effects were reported.

3.5.2 *Nail Psoriasis*

Nail psoriasis has been reported in up to 50% of patients with psoriasis. Nails in general are difficult to treat and respond slowly. Up to now, there has been no consistently effective treatment for psoriatic nails. Lately, it was shown that calcipotriol ointment is effective in the treatment of nail psoriasis (Petrow 1995).

3.5.3 *Face and Flexures*

Even if the use of calcipotriol ointment is not recommended on face and flexures due to its irritant nature, most patients tolerate vitamin D analogues on these sites. Lately, the tolerability and efficacy of calcitriol 3 μg/g and tacrolimus 0.3 mg/g ointment in chronic plaque psoriasis affecting facial and genitofemoral regions was analyzed (Liao et al. 2007). In this double-blind, parallel, 6-week study, 50 patients were randomized in a 1:1 ratio to apply calcitriol or tacrolimus twice a day. The primary efficacy variable was the mean reduction of the target area score (TAS); the secondary efficacy variable was the percentage of patients with a Physician's Global Assessment (PGA) score of 5 (clear) and 4 (almost clear) at the end of the study. Both calcitriol and tacrolimus were well tolerated. Although calcitriol induced perilesional erythema in a statistically significant higher proportion of patients than tacrolimus (55% vs. 16% at week 6; $P<0.05$), it did not necessitate treatment discontinuation. As a result of the study, tacrolimus was significantly more effective than calcitriol based on a significant reduction of mean TAS (67% vs. 51%; $P<0.05$) as well as more patients achieving complete or almost complete clearance by the PGA (60% vs. 33%; $P<0.05$). The authors concluded that both calcitriol 3 μg/g and tacrolimus 0.3 mg/g are safe and well-tolerated therapeutic agents for treating psoriasis in sensitive areas (Liao et al. 2007). Tacrolimus demonstrated a slightly more effective clinical outcome than calcitriol (Liao et al. 2007).

3.5.4 *Skin Lesions in Children*

During the last few years it has been proved that topical application of calcitriol ointment (3 μg of calcitriol per gram of petrolatum) is an effective, safe, reliable therapy to cure psoriatic skin lesions in children (Saggese et al. 1993; Perez et al. 1995).

3.5.5 *Psoriatic Lesions in HIV Patients*

Holick treated an HIV-positive patient suffering from psoriatic skin lesions with topical and oral calcitriol (Holick 1991). The patient responded well, and there was no evidence that it had enhanced HIV disease activity or caused alterations in the number of T lymphocytes or CD4+ and CD8+ cells. Other case reports also demonstrated the efficacy and safety of vitamin D analogues in the treatment of psoriasis (Gray et al. 1992).

3.5.6 Combination of Vitamin D Analogues with Other Therapies

Lately, it has been reported that the efficacy of topical treatment with vitamin D analogues against psoriasis can be augmented by combining it with other therapies, including very low dose oral cyclosporine (2 mg/kg/day), oral acitretin, topical dithranol, topical steroids, and UVB or narrow-band UVB phototherapy (Grossman et al. 1994; Kerscher et al. 1993; Cambazard et al. 1996; Ortonne 1994; Kragballe 1990). Complete clearing or 90% improvement in PASI was noted in 50% of patients treated with calcipotriol/cyclosporine versus 11.8% in the placebo/cyclosporine group. No difference was found between the groups concerning side effects in this study.

Addition of calcipotriol ointment to oral application of acitretin (a vitamin A analogue) was demonstrated to produce a significantly better treatment response achieved with a lower cumulative dose of acitretin in patients with severe extensive psoriasis vulgaris compared with the group of patients treated with oral acitretin alone. The number of patients reporting adverse events was similar for the two treatment groups (Cambazard et al. 1996).

Combined topical treatment with calcipotriol ointment (50 μg/g) and betamethasone ointment was recently shown to be slightly more effective and caused less skin irritation than calcipotriol used twice a day (Ortonne 1994). Consequently, calcipotriol is now also available in a combined formulation with betamethasone dipropionate. The safety and efficacy of the combined formulation of calcipotriol and betamethasone dipropionate when used over a 4-week period is well documented. Recently, several publications report on the safety of this product when used for 52 weeks, representing an option for maintenance therapy in psoriatic patients (Toole 2007). In a recent investigation using an economic model to simulate the costs and benefits (Bottomley et al. 2007), the cost-effectiveness of the two-compound formulation (calcipotriol and betamethasone dipropionate) was compared with other topical treatments commonly used in the management of moderately severe plaque psoriasis in Scotland. In that study, the calcipotriol and betamethasone dipropionate formulation was associated with reduced costs and superior outcomes when compared to other topical treatments.

Kragballe et al. reported that the efficacy of topical calcipotriol treatment for psoriasis can be improved by simultaneous UVB phototherapy. Lately, therapy of psoriasis with combined topical calcipotriol and narrow-band UVB has been shown to be highly effective for treating psoriatic plaques (Kragballe 1990). Experimental investigations also demonstrated that calcipotriol is degraded by UV radiation and suggested that calcipotriol be applied after phototherapy but not immediately before. Nevertheless, it was reported that the clinical effect of vitamin D analogues is not inactivated by subsequent UV exposure (Adachi et al. 2008). In that experiment, calcipotriol or maxacalcitol ointment was topically applied to psoriatic plaques of six patients immediately before or after phototherapy on the right or left side of the body, respectively. The topical application of vitamin D_3 analogues either before or after irradiation by psoralen and UVA radiation (PUVA) or narrow-band (NB)-UVB showed similar effects in all patients. The authors concluded that the therapeutic effects of vitamin D analogues were not clinically inactivated by subsequent irradiation with PUVA or NB-UVB phototherapy.

3.5.7 *Other Skin Disorders with Vitamin D Analogues*

Earlier in this century, vitamin D_3 was used in dermatology in large pharmacological doses for the treatment of scleroderma, psoriasis, lupus vulgaris, and atopic dermatitis. These first attempts of vitamin D treatment in dermatology were rapidly abandoned because of severe vitamin D intoxication that caused hypercalcemia, hypercalciuria, and kidney stones and because other new treatments were introduced for the therapy of these diseases.

3.5.7.1 Vitamin D and Ichthyosis

A double-blind, bilaterally paired, comparative study showed the effectiveness of topical treatment with calcipotriol ointment on congenital ichthyoses (Lucker et al. 1994). Reduced scaling and roughness on the calcipotriol-treated side was shown in all patients with lamellar ichthyosis and bullous ichthyotic erythroderma of Brocq. The only patient with Comel-Netherton syndrome was treated and showed mild improvement, and the only patient suffering from ichthyosis bullosa of Siemens who was treated with calcipotriol did not show any change in severity on the calcipotriol-treated side compared to the vehicle-treated side.

3.5.7.2 Vitamin D and Scleroderma

Some case reports point at the efficacy of vitamin D analogues for the treatment of scleroderma. Humbert et al. (1993) reported that oral administration of calcitriol 1.0–2.5 μg/day alleviates skin involvement, probably by inhibiting fibroblast proliferation and dermal collagen deposition.

3.5.7.3 Vitamin D and Vitiligo

A great variety of therapeutic agents are being used for the treatment of vitiligo, but treatment remains a challenge. Recent investigations indicate that vitamin D analogues may be efficient and safe in the treatment of vitiligo. Some studies report successful therapy with topically applied vitamin D analogues such as calcipotriol (Kumaran et al. 2006) or tacalcitol alone, and others report successful therapy with topically applied vitamin D analogues in combination with UV (Amano et al. 2008) or laser (Goldinger et al. 2007) therapy.

3.5.7.4 Vitamin D and Skin Cancer

In vitro studies have demonstrated strong antiproliferative and prodifferentiating effects of vitamin D analogues in many VDR-expressing tumor cell lines, including malignant melanoma, SCC, and leukemic cells (Texereau and Viac 1992; Koeffler et al. 1985). In vivo studies have supported these results and showed that active vitamin D analogues block proliferation and tumor progression of epithelial tumors in rats (Colston et al. 1992). Furthermore, it was shown that administration of calcitriol reduced the number of lung metastases after implantation of lung carcinoma cells in mice (Franceschi et al. 1987). Inhibition of tumor growth of human malignant melanoma and colonic cancer xenografts was also mice, but only at high doses of calcitriol (Eisman et al. 1987). Little is known concerning the effects of calcitriol on the formation of metastases in patients with malignant melanoma or SCC of the skin.

3.5.7.5 Vitamin D and Other Skin Diseases

A number of case reports have demonstrated positive effects of topical treatment with vitamin D analogues in a variety of skin diseases, such as transient acantholytic dermatosis (Grover's disease), inflammatory linear verrucous epidermal nevus (ILVEN), disseminated superficial actinic porokeratosis, pityriasis rubra pilaris, epidermolytic palmoplantar keratoderma of Vorner, and Sjögren-Larsson syndrome. These promising observations will have to be further evaluated in clinical trials.

Take-Home Messages

- The safety and efficacy of calcitriol and analogues for the topical treatment of psoriasis has been extensively shown, and calcitriol and analogues are at present a first-choice treatment for psoriasis in many countries.
- An increasing number of studies, reviewed in this chapter, demonstrate positive effects of vitamin D or its analogues in a variety of other skin diseases, such as ichthyosis, scleroderma, and vitiligo; moreover, vitamin D and analogues may be effective in the prevention of skin cancer.
- Studies with oral administration demonstrate benefits of vitamin D and analogues in the treatment of psoriasis and other skin diseases, but this use is limited by the possibility of hypercalcemia and other systemic side effects.
- Vitamin D and analogues exert potent immunomodulatory effects in the skin and in other tissues, making vitamin D compounds promising candidates for the treatment of inflammatory skin diseases, including atopic dermatitis.

- In vitro studies have demonstrated strong antiproliferative and prodifferentiating effects of vitamin D analogues in many VDR-expressing tumor cell lines, including malignant melanoma, cutaneous SCC, and leukemic cells, making vitamin D compounds promising candidates for the treatment of malignancies, including skin cancer.
- All of the beneficial effects of topical and perhaps oral administration of vitamin D for treatment of psoriasis and other skin diseases will have to be further evaluated in clinical trials, which are increasing dramatically with time.

References

Adachi Y, Uchida N, Matsuo T, Horio T. Clinical effect of vitamin D3 analogues is not inactivated by subsequent UV exposure. Photodermatol Photoimmunol Photomed. 2008;24(1):16–8.

Alam M, Ratner D. Cutaneous squamous-cell carcinoma. N Engl J Med. 2001;344:975–83.

Amano H, Abe M, Ishikawa O. First case report of topical tacalcitol for vitiligo repigmentation. Pediatr Dermatol. 2008;25(2):262–4.

Bielawska A, Linardic CM, Hannun YA. Modulation of cell growth and differentiation by ceramide. FEBS Lett. 1992;307:211–4.

Bikle DD, Neumanic MK, Gee E, Elias P. 1, 25-Dihydroxyvitamin D_3 production by human keratinocytes. J Clin Invest. 1986;78:557–66.

Bikle DD, Halloran BP, Riviere JE. Production of 1, 25-Dihydroxyvitamin D_3 by perfused pig skin. J Clin Invest. 1994;102:796–8.

Bottomley JM, Auland ME, Morais J, Boyd G, Douglas WS. Cost-effectiveness of the two-compound formulation calcipotriol and betamethasone dipropionate compared with commonly used topical treatments in the management of moderately severe plaque psoriasis in Scotland. Curr Med Res Opin. 2007;23(8):1887–901.

Cambazard, van de Kerkhof PCM, Hutchinson PE, and the Calcipotriol Study Group. 1996. Proceedings of the 3 rd International Calcipotriol Symposium, Munich, Germany, 23 March 1996.

Colston KW, Chander SK, Mackay AG, Coombes RC. Effects of synthetic vitamin D analogues on breast cancer cell proliferation in vivo and in vitro. Biochem Pharmacol. 1992;44:693–702.

De Haes P, Garmyn M, Degreef H, Vantieghem K, Bouillon R, Segaert S. 1, 25-dihydroxyvitamin D_3 inhibits ultraviolet B induced apoptosis, jun kinase activation, and interleukin-6 production in primary human keratinocytes. J Cell Biochem. 2003;89:663–73.

De Haes P, Garmyn M, Verstuyf A, De Clercq P, Vandewalle M, Vantieghem K, et al. Two 14-epi analogues of 1, 25-dihydroxyvitamin D_3 protect human keratinocytes against the effects of UVB. Arch Dermatol Res. 2004;295:527–34.

De Haes P, Garmyn M, Verstuyf A, De Clercq P, Vandewalle M, Degreef H, Vantieghem K, Bouillon R, Segaert S. 1, $25(OH)_2D_3$ and analogues protect primary human keatinocytes against UVB-induced DNA damage. 2005.

De Korte J, van der Valk PG, Sprangers MA, Damstra RJ, Kunkeler AC, Lijnen RL, et al. A comparison of twice-daily calcipotriol ointment with once-daily short-contact dithranol cream therapy: quality-of-life outcomes of a randomized controlled trial of supervised treatment of psoriasis in a day-care setting. Br J Dermatol. 2008;158(2):375–81.

Eisman JA, Barkla DH, Tutton PJM. Suppression of in vivo growth of human cancer solid tumor xenografts by 1α,25-dihydroxyvitamin D_3. Cancer Res. 1987;47:21–5.

Elwood JM, Jopson J. Melanoma and sun exposure: an overview of published studies. Int J Cancer. 1997;73:198–203.

Ezquerra GM, Regana MS, Millet PU. Combination of acitretin and oral calcitriol for treatment of plaque-type psoriasis. Acta Derm Venereol. 2007;87(5):449–50.

Franceschi RT, Linson CJ, Peter CT, Romano PR. Regulation of cellular adhesion and fibronectin synthesis by 1α,25-dihydroxyvitamin D_3. J Biol Chem. 1987;262:4165–71.

Fu GK, Lin D, Zhang MY, Bikle DD, Shackleton CH, Miller WL, et al. Cloning of human 25-hydroxyvitamin D-1 alpha-hydroxylase and mutations causing vitamin D-dependent rickets type 1. Mol Endocrinol. 1997;11:1961–70.

Geilen CC, Bektas M, Wieder T, Orfanos CR. 1α,25-Dihydroxyvitamin D_3 induces sphingomyelin hydrolysis in HaCaT cells via tumor necrosis factor α. J Biol Chem. 1997;272:8997–9001.

Gniadecki R. Stimulation versus inhibition of keratinocyte growth by 1,25-Dihydroxyvitamin D3: dependence on cell culture conditions. J Invest Dermatol. 1996;106(3):510–6.

Goldinger SM, Dummer R, Schmid P, Burg G, Seifert B, Läuchli S. Combination of 308-nm xenon chloride excimer laser and topical calcipotriol in vitiligo. J Eur Acad Dermatol Venereol. 2007;21(4):504–8.

Gombard HF, Borregaard N, Koeffler HP. Human cathelicidin antimicrobial peptide (CAMP) gene is a direct target of the vitamin D receptor and is strongly up-regulated in myeloid cells by 1,25-dihydroxyvitamin D3. FASEB J. 2005;19(9):1067–77.

Gray JD, Bottomley W, Layton AM, Cotterill JA, Monteiro E. The use of calcipotriol in HIV-related psoriasis. Clin Exp Dermatol. 1992;17(5):342–3.

Green C, Ganpule M, Harris D, Kavanagh G, Kennedy C, Mallett R, et al. Comparative effects of calcipotriol (MC 903) solution and placebo (vehicle of MC 903) in the treatment of psoriasis of the scalp. Br J Dermatol. 1994;130:483–7.

Griffin M, Kumar R. Effects of 1α,25-Dihydroxyvitamin D_3 and its analogs on dendritic cell function. J Cell Biochem. 2003;88:323–6.

Grossman RM, Thivolet J, Claudy A, Souteyrand P, Guilhou JJ, Thomas P, et al. A novel therapeutic approach to psoriasis with combination calcipotriol ointment and very low-dose cyclosporine: a result of a multicenter placebo-controlled study. J Am Acad Dermatol. 1994;31:68–74.

Gupta R, Dixon KM, Deo SS, Holliday CJ, Slater M, Halliday GM, et al. Photoprotection by 1, 25-Dihydroxyvitamin D_3 is associated by an increase in p53 and a decrease in nitric oxide products. J Invest Dermatol. 2006;127:707–15.

Hanada K, Sawamura D, Nakano H, Hashimoto I. Possible role of 1,25-dihydroxyvitamin D_3–induced metallothionein in photoprotection against UVB injury in mouse skin and cultured rat keratinocytes. J Dermatol Sci. 1995;9:203–8.

Helfrich YR, Kang S, Hamilton TA, Voorhees JJ. Topical becocalcidiol for the treatment of psoriasis vulgaris: a randomized, placebo-controlled, double-blind, multicentre study. Br J Dermatol. 2007;157(2):369–74.

Holick MF. Photobiology, physiology and clinical applications for vitamin D. In: Goldsmith LA, editor. Physiology, biochemistry and molecular biology of the skin. 2nd ed. New York: Oxford University Press; 1991. p. 928–56.

Holick MF. Evolution and function of vitamin D. Recent Results Cancer Res. 2003;164:3–28.

Holick MF. Vitamin D deficiency. NEJM. 2007;357:266–81.

Holick MF, Reichrath J. Clinical utility of 1,25-dihydroxyvitamin D_3 and its analogues for the treatment of psoriasis. In: Holick MF, editor. Vitamin D: physiology, molecular biologic and clinical aspects. Totowa: The Humana Press Inc; 1999. p. S357–73.

Holick MF, Chen ML, Kong XF, Sanan DK. Clinical uses for calciotropic hormones 1,25-dihydroxyvitamin D_3 and parathyroid hormone related peptide in dermatology: a new perspective. J Invest Dermatol Symp Proc. 1996;1:1–9.

Humbert P, Dupond JL, Agache P, Laurent R, Rochefort A, Drobacheff C, et al. Treatment of scleroderma with oral 1,25-dihydroxyvitamin D_3: evaluation of skin involvement using non-invasive techniques. Results of an open prospective trial. Acta Derm Venereol Stockh. 1993;73:449–51.

Katayama I, Minatohara K, Yokozeki H, Nishioka K. Topical vitamin D_3 downregulates IgE-mediated murine biphasic cutaneous reactions. Int Arch Allergy Immunol 1996;111(1):71–6.

Kerscher M, Volkenandt M, Plewig G, Lehmann P. Combination phototherapy of psoriasis with calcipotriol and narrow band UVB. Lancet. 1993;342:923.

Koeffler HP, Hirji K, Itri L. 1,25-dihydroxyvitamin D_3: in vivo and in vitro effects on human preleukemic and leukemic cells. Cancer Treat Rep. 1985;69:1399–407.

Kragballe K. Combination of topical calcipotriol (MC 903) and UVB radiation for psoriasis vulgaris. Dermatologica. 1990;181:211–4.

Kragballe K, Beck HI, Sogaard H. Improvement of psoriasis by topical vitamin D_3 analogue (MC 903) in a double-blind study. Br J Dermatol. 1988;119:223–30.

Kragballe K, Gjertsen BT, de Hoop D, Karlsmark T, van de Kerhof PCM, Larko O, et al. Double-blind right/left comparison of calcipotriol and betametasone valerate in treatment of psoriasis vulgaris. Lancet. 1991;337:193–6.

Kragballe K, Barnes L, Hamberg KJ, Hutchinson P, Murphy F, Moller S, et al. Calcipotriol cream with or without concurrent topical corticosteroid in psoriasis: Tolerability and efficacy. Br J Dermatol. 1998;139:649–54.

Kricker A, Armstrong BK, English DR. Sun exposure and non-melanocytic skin cancer. Cancer Causes Control. 1994;5:367–92.

Kumaran MS, Kaur I, Kumar B. Effect of topical calcipotriol, betamethasone dipropionate and their combination in the treatment of localized vitiligo. J Eur Acad Dermatol Venereol. 2006;20(3):269–73.

Lee JH, Youn JI. The photoprotective effect of 1, 25-Dihydroxyvitamin D_3 on ultraviolet B-induced damage in keratinocyte and its mechanisms of action. J Dermatol Sci. 1998;18:11–8.

Lehmann B, Tiebel O, Meurer M. Expression of vitamin D3 25-hydroxylase (CYP27) mRNA after induction by vitamin D3 or UVB radiation in keratinocytes of human skin equivalents-a preliminary study. Arch Dermatol Res. 1999;291:507–10.

Lehmann B, Genehr T, Knuschke P, Pietsch J, Meurer M. UVB-induced conversion of 7-dehydrocholesterol to 1alpha25-dihydroxyvitamin D_3 in an in vitro human skin equivalent model. J Invest Dermatol. 2001;117:1179–85.

Liao YH, Chiu HC, Tseng YS, Tsai TF. Comparison of cutaneous tolerance and efficacy of calcitriol 3 microg g(−1) ointment and tacrolimus 0.3 mg g(−1) ointment in chronic plaque psoriasis involving facial or genitofemoral areas: a double-blind, randomized controlled trial. Br J Dermatol. 2007;157(5):1005–12.

Lucker GP, van de Kerkhof PC, van Dijk MR, Steijlen PM. Effect of topical calcipotriol on congenital ichthyosis. Br J Dermatol. 1994;131:546–50.

MacLaughlin JA, Gange W, Taylor D, Smith E, Holick MF. Cultured psoriatic fibroblasts from involved and uninvolved sites have partial but not absolute resistance to the proliferation-inhibition activity of 1,25-dihydroxyvitamin D_3. Proc Natl Acad Sci USA. 1985;82:5409–12.

Manggau M, Kim DS, Ruwisch L, Vogler R, Korting HC. 1α,25-Dihydroxyvitamin D_3 protects human keratinocytes from apoptosis by the formation of sphingosine-1-phosphate. J Invest Dermatol. 2001;117:1241–9.

Morimoto S, Kumahara Y. A patient with psoriasis cured by 1α-hydroxyvitamin D_3. Med J Osaka Univ. 1985;35(3–4):51–4.

Morimoto S, Yochikawa K, Kozuka T, Kitano Y, Imawaka S, Fukuo K, Koh E, Kumahara Y. An open study of vitamin O_3 treatment in psoriasis vulgaris. Br J Dermatol 1986;115:421–9.

Okasaki T, Bell RM, Hannun YA. Sphingomyelin turnover induced by vitamin D_3 in HL-60 cells. Role in cell differentiation. J Biol Chem. 1989;264:19076–80.

Ortonne JP. Calcipotriol in combination with betametasone diproprionate. Nouv Dermatol. 1994;13:736–51.

Osterlind A, Tucker MA, Stone BJ, Jensen OM. The Danish case-control study of cutaneous malignant melanoma. II Importance of UV-light exposure. Int J Cancer. 1988;42:319–24.

Perez A, Chen TC, Turner A, Holick MF. Pilot study of topical calcitriol (1,25-dihydroxyvitamin D_3) for treating psoriasis in children. Arch Dermatol. 1995;131:961–2.

Perez A, Chen TC, Turner A, Raab R, Bhawan J, Poche P, et al. Efficacy and safety of topical calcitriol (1,25-dihydroxyvitamin D_3) for the treatment of psoriasis. Br J Dermatol. 1996a;134:238–46.

Perez A, Raab R, Chen TC, Turner A, Holick MF. Safety and efficacy of oral calcitriol (1,25-dihydroxyvitamin D_3) for the treatment of psoriasis. Br J Dermatol. 1996b;134:1070–8.
Petrow W. Treatment of a nail psoriasis with calcipotriol. Akt Dermatol. 1995;21:396–400.
Reichrath J. The challenge resulting from positive and negative effects of sunlight: how much solar UV-exposure is appropriate to balance between risks of vitamin D deficiency and skin cancer? Prog Biophys Mol Biol. 2006;92:9–16.
Reichrath J, Müller SM, Kerber A, Baum HP, Bahmer FA. Biologic effects of topical calcipotriol (MC 903) treatment in psoriatic skin. J Am Acad Dermatol. 1997a;36:19–28.
Reichrath J, Perez A, Chen TC, Kerber A, Bahmer FA, Holick MF. The effectiveness of topical 1,25-dihydroxyvitamin D_3 (1,25$(OH)_2D_3$) application in the treatment of psoriasis: an immunohistological evaluation. Acta Derm Venereol. 1997b;77:268–72.
Saggese G, Federico G, Battini R. Topical application of 1,25 dihydroxyvitamin D_3 (calcitriol) is an effective and reliable therapy to cure skin lesions in psoriatic children. Eur J Pediatr. 1993;152:389–92.
Serup J. Calcipotriol irritation: mechanism, diagnosis and clinical implication. Acta Derm Venereol (Stockh) Abstract. 1994;186:42S.
Sigmundsdottir H, Pan J, Debes GF, Alt C, Habtezion A, Soler D, Butcher EC. DCS metabolize sunlight–induced vitamin D_3 to "program" T cell attraction to the epidermal chemokine CCL27. Nat Immunol 2007;8(3):285–93.
Solvoll K, Soyland E, Sandstad B, Drevon CA. Dietary habits among patients with atopic dermatitis. Eur J Clin Nutr. 2000;54(2):93–7.
Texereau M, Viac J. Vitamin D, immune system and skin. Eur J Dermatol. 1992;2:258–64.
Toole JW. Calcipotriol and betamethasone dipropionate for the treatment of psoriasis: a 52-week study. Skin Therapy Lett. 2007;12(4):1–3.
Trémezaygues L. In vitro Untersuchungen in humanen Keratinozyten über mögliche protektive Eigenschaften von 1,25-dihydroxyvitamin D_3 gegen UVB- und niedrigdosierte ionisierende Strahlung. 2009.
Tzaneva S, Hönigsmann H, Tanew A. Observer-blind, randomized, intrapatient comparison of a novel 1% coal tar preparation (Exorex R) and calcipotriol cream in the treatment of plaque type psoriasis. Br J Dermatol. 2003;149:350–3.
Van de Kerkhof PC, Vissers WH. The topical treatment of psoriasis. Skin Pharmacol Appl Skin Physiol. 2003;16:69–83.
van de Kerkhof PCM, van Bokhoven M, Zultak M, Czarnetzki BM. A double-blind study of topical 1α-25-dihydroxyvitamin D_3 in psoriasis. Br J Dermatol. 1989;120:661–4.
Weber G, Heilborn JD, Chamorro Jimenez CI, Hammarsjö A, Törmä H, Ståhle M. Vitamin D induces the antimicrobial protein hCAP18 in human skin. J Invest Dermatol. 2005;124(5):1080–2.
Zhu X, Wang B, Zhao G, Gu J, Chen Z, Briantais P, et al. An investigator-masked comparison of the efficacy and safety of twice daily applications of calcitriol 3 microg/g ointment vs. calcipotriol 50 microg/g ointment in subjects with mild to moderate chronic plaque-type psoriasis. J Eur Acad Dermatol Venereol. 2007;21(4):466–72.

Chapter 4
Photoprotection of the Skin with Vitamins C and E: Antioxidants and Synergies

Karen E. Burke

Core Messages

- Sunscreen alone is not sufficient to provide optimal protection from ultraviolet-induced and other environmental free-radical damage to the skin.
- The antioxidants vitamins C and E have been extensively researched and proven to protect the skin against photodamage (sunburn, tanning, photoaging, precancers, cancer).
- When formulated correctly and at a high enough concentration, vitamins C and E are more protective when applied topically to the skin than if taken orally.
- An effective formulation requires the natural molecular form of the antioxidant (the nonesterified, active isomer) in a composition that maintains stability and delivers enough active antioxidant to the deep layers of the skin: Vitamin C must be ascorbic acid, optimally at a concentration of 15–20%, and vitamin E must be d-α-tocopherol, optimally at a concentration of 2–5%.

4.1 Introduction

Our skin is the largest organ of our body and the organ most exposed to the environment. Unfortunately, our skin suffers from ultraviolet A (UVA) and B (UVB) exposure as well as from urban pollutants. Research has recently demonstrated a synergistic enhancement of oxidative damage to the skin when the skin is exposed to UVA

K.E. Burke (✉)
Department of Dermatology, Mount Sinai Medical Center, One Gustave Levy Place, River Court, 429 East 52nd Street, New York, NY 10029, USA
e-mail: kebmdphd@gmail.com

A. Pappas (ed.), *Nutrition and Skin: Lessons for Anti-Aging, Beauty and Healthy Skin*,
DOI 10.1007/978-1-4419-7967-4_4, © Springer Science+Business Media, LLC 2011

and environmental irritants, including cigarette smoke (Burke and Wei 2009; Wang et al. 2003). This cumulative damage leads to unattractive premature aging of the skin and to precancers and cancers of the skin.

One in five Americans develop at least one skin cancer in his or her lifetime, and one American dies each hour of skin cancer (The Skin Cancer Foundation – Guidelines 2010). Also devastating is the fact that one in four potentially lethal melanomas occurs in individuals younger than 40 years of age (The Skin Cancer Foundation – Guidelines 2010). Especially because we enjoy more leisure time outside not only in the summer (where UV exposure is increased by reflection from sea and sand as well as concrete playgrounds) but also in the winter (when we can go south for bursts of extra UV or north for skiing at high altitudes with increased UV due to reflection from snow), the incidence of skin cancer is still rising.

Furthermore, we will all live longer than generations before, and we have the possibility of far better health at every age. We want to "look as young as we feel," not compromised by unattractive photoaging with mottled dark spots, wrinkles, and sagging skin.

Because more than 90% of skin cancers and all photoaging are caused by sun exposure, effective sun protection can prevent this damage. Unfortunately, fewer than 33% of the population apply sunscreen regularly despite the current increased publicity promoting the need for protection (The Skin Cancer Foundation – Guidelines 2010). Especially now that we have far better sunscreens in recent years than previously—with higher sun protection factors (SPFs), protection from UVA as well as UVB, noncomedogenic formulations that are cosmetically appealing (nonopaque, not greasy, often with attractive tint), and easy to apply (sprays, gels)—hopefully everyone will develop the habit of applying them frequently.

However, sunscreens alone are not enough to provide optimal protection for several reasons. First, even when applied properly at 2 mg/cm^2, broad-spectrum sunscreens decrease free radical damage by only 55% (Haywood et al. 2003). Second, even when conscientious individuals apply sunscreen generously, they can usually apply only one-fourth of the amount required to attain the full SPF. Therefore, a sunscreen labeled "SPF 30" gives only an effective SPF of 2.3 (Wulf et al. 1997). Furthermore, sunscreen must be applied generously to *all* exposed skin; balding men often miss their scalps, and most people miss around the eyes and mouth, in front of the ears, the neck, and the fingertips. Also, sunscreen must be applied every 1.5 h when outside or even when inside near a window (through which UVA does penetrate).

These limitations can be overcome by supplementation with both oral and topical antioxidants as an adjunct to sunscreen. Although many antioxidants are being studied for photoprotection, vitamins C and E have been the most extensively researched and have been proven effective. Each of these antioxidants does protect against ongoing photodamage when taken orally. However, it is impossible to attain protective levels of antioxidants through diet alone, so they must be taken as supplements. The Federal Drug Administration (FDA) has specified required daily amounts (RDA) of all vitamins to prevent deficiency diseases such as scurvy. The RDA of vitamin C is only 65 mg/day and of vitamin E 30 IU/day. However, to achieve optimal health benefits, including photoprotection (especially in older individuals), far higher intakes are recommended. Supplements of vitamin C at 1,000–3,000 mg/day and of d-α-tocopherol (natural vitamin E) at 400 IU/day are needed to attain significant photoprotection—equivalent to 100 oranges and 44 tablespoons of sunflower oil (5,275 cal), respectively, per day. For optimal protection against sunburn, tanning, and skin precancers and cancers, vitamin C must be taken for at least 5 days and vitamin E for at least 3–4 weeks before exposure and continued daily, especially during summer months.

Although oral vitamins C and E do protect against UV radiation and other free-radical damage to the skin, topical application is far more effective and gives the further advantage of reversing previously incurred photodamage. Far higher concentrations can be attained in the skin by topical application of the correct formulations: The skin level of vitamin C can be 27–40 times higher (Darr et al. 1992) and vitamin E 11 times higher (Burke et al. 2000) than with even the large orally ingested amounts recommended above. The challenge is to create stable formulations that give effective transcutaneous absorption of the active form, as discussed in detail in the following sections. Once-daily application of effective formulations containing vitamin C and vitamin E is adequate to provide a reservoir to the whole skin for protection against UV-induced sunburn, tanning, photoaging, and skin cancer as well as against all other free-radical damage. This application over time allows repair and actually acts to reverse prior photodamage.

4.2 Vitamin C

Vitamin C (L-ascorbic acid) is the body's major aqueous-phase antioxidant and is vital for life. Humans and other primates (and the guinea pig and Indian bat) cannot synthesize their own vitamin C, as do most other animals, including insects. In fact, animals synthesize very large quantities of vitamin C. (A 130-lb goat manufactures 13 g of vitamin C daily, equal to 200 times the FDA human requirement (Puling 1987). Furthermore, animals can make up to ten times this normal amount when under stress (Puling 1987), suggesting the importance of vitamin C supplementation to human physiology.

Because our skin is the organ that protects us from environmental free-radical stress, the presence of vitamin C is extremely important. Exposure to sunlight and environmental pollution depletes vitamin C from the center layers of the skin. Even minimal UV exposure of 1.6 minimal erythema dose (MED) decreases the level of vitamin C to 70% of the normal level, and exposure to 10 MED decreases the vitamin C to only 54% (Shindo et al. 1994). Exposure to ozone at a dose of 10 parts per million in city pollution decreases the level of epidermal vitamin C by 55% (Thiele et al. 1997).

Active L-ascorbic acid is such an excellent antioxidant that it is inherently unstable, turning brown as it is oxidized to dihydroascorbic acid when exposed to air. Therefore, the shelf life of most formulations containing pure vitamin C is short,

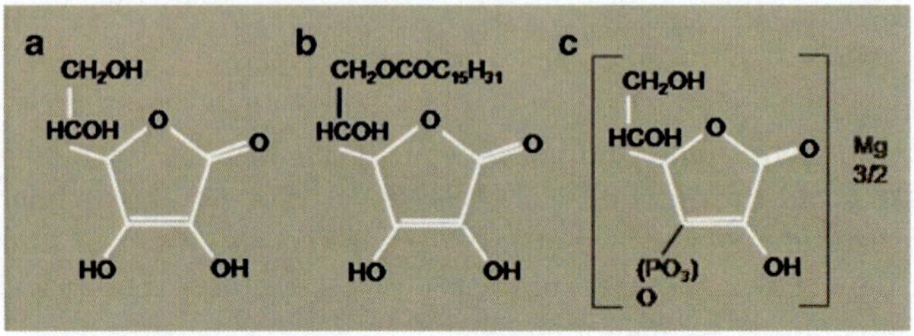

Fig. 4.1 Molecular structure of (**a**) L-ascorbic acid, (**b**) L-ascorbyl-6-palmitate, and (**c**) magnesium ascorbyl phosphate (Courtesy of Sheldon R. Pinnell, MD. Printed with his permission)

so other forms of vitamin C are usually used for topical application in lotions, creams, serums, and patches to overcome the problem of instability. However, these more stable esterified derivatives (ascorbyl-6-palmitate and magnesium ascorbyl phosphate whose structures are shown in Fig. 4.1) are not well absorbed (Darr et al. 1992) and are only minimally metabolized by the skin to the active, free acid form. To achieve photoprotection and other benefits to the skin with topical vitamin C, the formulation must contain L-ascorbic acid in a high enough concentration (at least 10%), be stable, and be at an acidic pH—less than the pKa (4.2) of vitamin C (Darr et al. 1992). (The optimal pH is 3.5.)

If these criteria are met, effective skin levels of vitamin C can be attained. Topical absorption has been proven by radioactive labeling studies in pigs (Darr et al. 1992). After treatment with 10% vitamin C cream, 8.2% was found in the dermis, and 0.7% was in the blood (Darr et al. 1992). Concentrations of 5%, 10%, 15%, 20%, or 25% vitamin C were tested: 20% resulted in the highest skin levels, with maximized concentration in the skin after 3 days of once-daily application (Pinnell et al. 2001). In these experiments, levels of vitamin C after topical application of 15% serum were shown to be a factor of about 27 times that which could ever be attained by even very high oral intake. If topical application is discontinued after skin saturation is achieved, high levels remain in the skin for more than 3 days (Pinnell et al. 2001).

Vitamin C has been proven to be photoprotective. Vitamin C does not absorb light in the UV spectrum, so vitamin C is itself not a sunscreen. However, as an antioxidant vitamin C deactivates UV-induced free radicals and decreases UVB erythema by 52% (Darr et al. 1992). This protection has been confirmed histologically: Treatment with topical 10% vitamin C decreased the number of abnormal "sunburn cells" by 40–60% (Darr et al. 1992) and reduced the UV damage to DNA 8-hydroxydeoxy guanosine (8-OHdG) by 62% (Darr et al. 1992).

Topical vitamin C is also directly antiinflammatory. Laser resurfacing (with the older CO_2 lasers) causes redness for at least 3–4 months after treatment. With vitamin C applied before and after laser surgery, this redness of inflammation was markedly decreased afterward, and healing took only 2 months (Alster and West 1998). Topical vitamin C can also be used effectively to treat the inflammation of rosacea (Bergfeld and Pinnell 1996). The mechanism of this antiinflammatory action has been researched in vitro with human cells in vitamin C-enriched media. Decreased activation of the transcription factor nuclear factor κβ (NF-κβ), the factor responsible for many preinflammatory cytokines such as tumor necrosis factor-α (TNFα), and interleukins IL-1, IL-6, and IL-8, was demonstrated (Carcamo et al. 2002).

By directly decreasing inflammation, postinflammatory hyperpigmentation can also be reduced. Vitamin C is itself an excellent depigmentating agent because it inhibits the enzyme tyrosinase (Maeda and Fukuda 1996), which is required for melanin production. I have also noted marked lightening of melasma and solar lentigines, even after only 2 months of daily application of topical vitamin C (15%).

Kameyama et al. (1996) demonstrated suppression of melanin formation by inhibiting tyrosinase in melanocytes and melanoma cells using 10% magnesium-L-ascorbyl-2 phosphate applied to human skin. Significant lightening of melasma and of lentigenes was observed in 19 of 34 patients (Kameyama et al. 1996).

Perhaps the most important action of vitamin C on the skin is direct stimulation of collagen synthesis. Vitamin C is an essential cofactor for the two enzymes required for collagen synthesis, prolyl hydroxylase (which makes the collagen molecule stable) and lysyl hydroxylase (which crosslinks the collagen to give structural strength) (Kivirikko and Myllyla 1996). Recent research has further demonstrated that vitamin C acts directly on DNA to increase the transcription rate and to stabilize the procollagen messenger RNA, thus regulating and maintaining the intercellular amount of collagen (Savini et al. 2002).

By enhancing collagen synthesis, vitamin C can directly correct the collagen loss that causes wrinkles. Exciting studies in vitro compared newborn with elderly (80- to 95-year-old) fibroblasts (Phillips et al. 1994). As shown in Fig. 4.2, elderly cells proliferated in vitro at only one-fifth the rate of newborn cells. However, when vitamin C was added to the culture medium, the elderly cells proliferated better than normal newborn fibroblasts. Even the newborn fibroblasts proliferated almost four times better when exposed to vitamin C (Phillips et al. 1994).

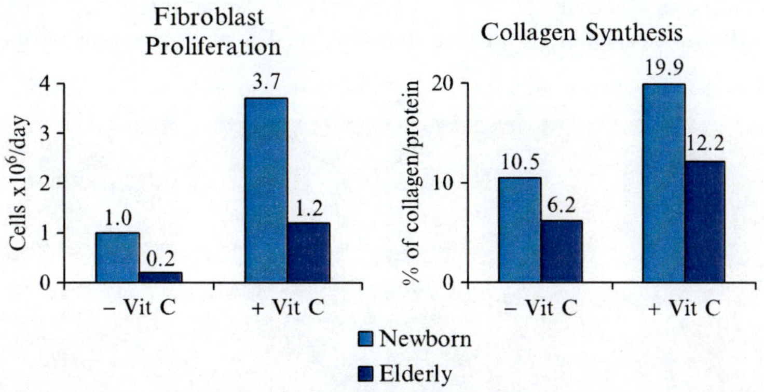

Fig. 4.2 Antiaging effects of vitamin C on newborn and elderly fibroblasts in vitro. Newborn fibroblasts (isolated from circumcised skin) and elderly fibroblasts (obtained from biopsies of individuals >90 years old) were grown in vitro with and without 10% L-ascorbic acid added to the culture medium. Fibroblast proliferation rate and synthesis of collagen per cell were measured (These graphs are representations of data from reference (Phillips et al. 1994))

Not only did fibroblasts increase proliferation in the presence of vitamin C, they also synthesized more collagen. Newborn fibroblasts synthesize a larger percentage of collagen than elderly cells; but, again, when elderly cells were exposed to vitamin C in vitro, they produced more collagen than the normal, newborn fibroblasts (Phillips et al. 1994). Surprisingly, the newborn cells also doubled the amount of collagen synthesized (Phillips et al. 1994). Thus, supplementing the skin with extra vitamin C not only combats the collagen destruction due to photodamage it can correct the loss (of about 1% per year) which occurs with natural, intrinsic aging (after approximately 45 years of age).

In contrast to the increased synthesis of collagen, other in vitro studies suggested that vitamin C may inhibit elastin biosynthesis by fibroblasts (Davidson et al. 1997). This might be advantageous in reducing the solar elastosis due to photodamage.

Topical vitamin C has also been shown to enhance collagen production in human skin in vivo (Humbert et al. 2003). Postmenopausal women who applied 5% vitamin C to one arm and half of the neck with placebo to the other side showed an increase in mRNA of collagens I and III (Humbert et al. 2003). Tissue levels of the inhibitor of metalloproteinase-1 (MMP-1) were also increased, thus decreasing UV-induced collagen breakdown. However, mRNA levels of elastin, fibrillin, and tissue inhibitor of MMP-2 remained unchanged. Clinically, a significant decrease was observed in deep furrows and substantiated by silicone replicas. Histology showed elastic tissue repair (Humbert et al. 2003). Other studies demonstrated a decrease in the crepey, laxity of forearm skin with restoration of a younger skinfold pattern after 6 months of once-daily treatment with 15% vitamin C.

All of these proven functions of topical vitamin C contribute to reversal of the appearance of photoaging: Photoprotection over many months allows the skin to correct previous photodamage; the synthesis of collagen and inhibition of MMP-I decreases wrinkles, (Humbert et al. 2003)and the inhibition of tyrosinase and the anti-inflammatory activity results in depigmenting solar lentigines (Maeda and Fukuda 1996). As seen in a split face study in Fig. 4.3, after 4 months of once-daily

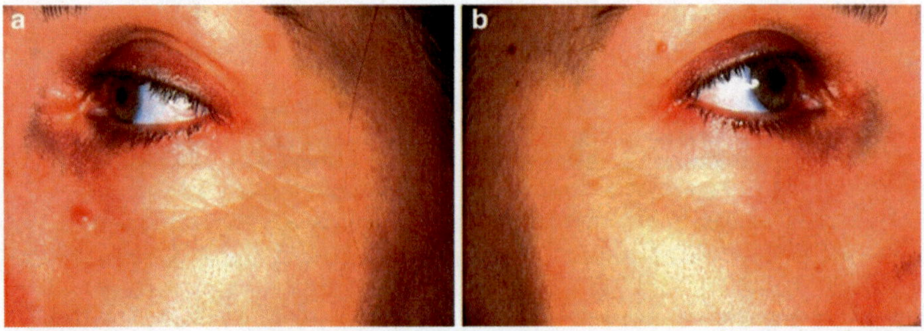

Fig. 4.3 A woman in her late twenties was treated once daily for 4 months (**a**) on the left side of the face with vehicle cream and (**b**) on the right side of the face with vitamin C (15%). The periorbital wrinkles on the right were markedly decreased by treatment with vitamin C

treatment with 15% topical vitamin C on the right and placebo cream on the left, the right periorbital wrinkles were clearly reduced, and the skin acquired a healthy, more youthful glow.

Topical vitamin C has also been shown to increase the synthesis of several specific sphingolipids of the skin surface (Uchida et al. 2001). With these lipids, vitamin C helps the natural moisturization of the skin as it enhances the protective barrier function (C. Catiel-Higournenc Ferrais, C. Guey, R. Schmidt, et al. L'Oréal Advanced Research Laboratories, Clichy and Aulnay-sous-Bois, France. Personal communication).

4.3 Vitamin E

Natural vitamin E is the most important lipid-soluble, membrane-bound antioxidant in the body. As a free-radical quencher, vitamin E deactivates these aggressive radicals and terminates damaging chain reactions, protecting primarily the fatty components of cell membranes. Vitamin E is synthesized only in plants, so all animals require a nutritional source. The highest levels in the body are in fatty tissue where vitamin E is stored. Vitamin E is delivered to the skin by sebum (Podda et al. 1996; Thiele 2001).

As the stratum corneum is the outermost defense of the body, this layer is the first to absorb the oxidative stress of sunlight and pollution. Vitamin E is depleted by this exposure to both UV (Shindo et al. 1994) and to the ozone of environmental pollution (Thiele et al. 1997), so its concentration is highest at the lower levels of the stratum corneum with a decreasing gradient outward. Thus, increasing the skin's surface concentration of vitamin E by topical application is especially important. The lipophilic structure of vitamin E enhances absorption and makes it cosmetically attractive for application as a moisturizer.

Several forms of vitamin E exist in natural dietary sources. The form found in mammalian tissues with by far the greatest biological activity is pure, nonesterified RRR-α-tocopherol (or d-α-tocopherol) (Mayer et al. 1993; Burton et al. 1998) with three methyl groups on the 6-chromal ring. Humans use predominantly α-tocopherol because a specific α-tocopherol transfer protein selectively transfers α-tocopherol into lipoproteins (Azzi et al. 2000). The other natural forms are β, γ, and δ, which contain only one or two methyl groups on the 6-chromal ring (as shown in Fig. 4.4). Relative to the α form, the β, γ, and δ RRR-tocopherols give only 42%, 72%, and 40%, respectively, of the protection against post-UV edema (Potokar et al. 1990).

Synthetic vitamin E is "dl" or "all-*rac*," a mixture of eight stereoisomers. These synthetic isomers are esterified (to acetates and succinates) for use in commercial vitamins and some topical formulations because the esters are far more stable. This ester must be hydrolyzed before there is any biological activity, a reaction that readily occurs in the stomach after oral ingestion or in cell and organ culture but is very slow after topical application. The skin has only a limited ability to cleave

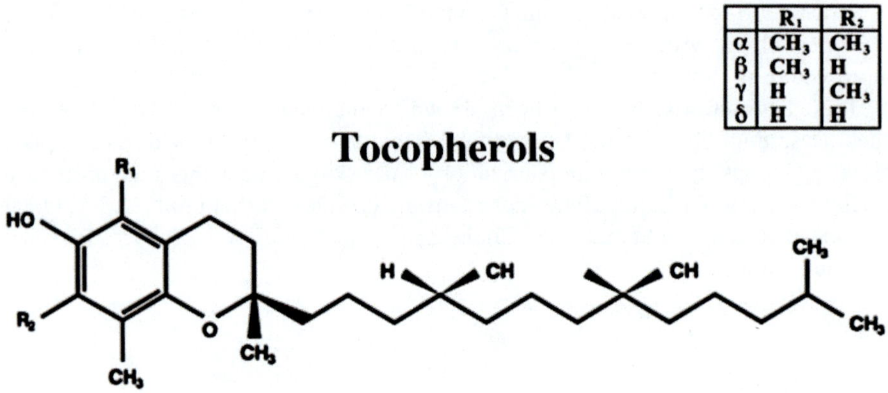

Fig. 4.4 Molecular structure of tocopherol isomers

the esterified forms of vitamin E to the active free tocopherol form, so the antioxidant potential of the esters is minimal (Gensler et al. 1996; van Hanegouwne et al. 1995). Furthermore, the all-*rac* form of vitamin E has been reported to cause allergic contact dermatitis (Hart 1990) and erythema multiforme (Saperstein et al. 1984) when applied topically. No such adverse reaction has been reported with d-α-tocopherol.

Many studies have demonstrated protection from UV-induced damage to the skin by applying various topical vitamin E formulations. Even various forms of topical vitamin E that are less metabolically potent when applied topically than the nonesterified d-α-tocopherol have demonstrated protection from the acute (Burke et al. 2000; Trevithick et al. 1992; Record et al. 1991; Marenus et al. 1990; Berton et al. 1998) UV-induced damage of inflammation (erythema, sunburn) and hyperpigmentation (tanning) as well as protection from the chronic UV-induced damage of actinic keratosis and skin cancer (Marenus et al. 1990; Berton et al. 1998; Bissett et al. 1992a, b; Gensler and Magdaleno 1991; Gerrish and Gensler 1993).

Few controlled studies have directly compared the efficacy of the various forms of topical vitamin E for photoprotection. Topical α-tocopheryl acetate was shown to be less effective than α-tocopherol in protecting against UV-induced erythema in rabbits (Gensler et al. 1996; Roshchupkin et al. 1979) and against UV-induced photoaging in mice (Bissett et al. 1990). In one mouse model, topical α-tocopheryl succinate and α-tocopheryl acetate not only failed to inhibit UVB-induced immunosuppression and carcinogenesis but appeared to enhance carcinogenesis (Gensler et al. 1996). In a 44-week mouse study (longer than most other published experiments), both d-α-tocopherol and d-α-tocopheryl succinate were proven effective in protecting against all acute and chronic UV-induced damage, with d-α-tocopherol most effective for all parameters (i.e., decreasing sunburn, tanning, skin cancer incidence) (Burke et al. 2000).

The results tabulated in Table 4.1 (Burke et al. 2000) show that topical d-α-tocopherol inhibited blistering sunburn by 73% and tanning by 40%. The topical ester d-α-tocopheryl succinate was more effective than oral vitamin E but less effective that the nonesterified topical d-α-tocopherol for both inhibition of acute

Table 4.1 Protection from acute UVB-induced damage with vitamin E

Treatment[a]	Blistering sunburn[b] (%)	Degree of tanning[c]
d-α-Tocopherol lotion (5%)	27	1.93
d-α-Tocopheryl succinate lotion (5%)	67	2.07
Oral d-α-tocopheryl acetate	73	2.43
Vehicle lotion	100	3.75

Data are from Burke et al. (2000)

[a]Skh:2 female mice were treated daily (5 days/week) for 2 weeks prior to UVB exposure and throughout the period of irradiation (three exposures per week), as described in detail in reference 7

[b]Percentage of mice (in each group of 15) that developed a blistering sunburn as the UV exposure was begun at low doses of 0.75 minimum erythema dose and increased sequentially to the final maintenance exposure time

[c]UV-induced pigmentation of each mouse after exposure to UV radiation was subjectively graded by two "blinded" investigators after 12 weeks. Values are averages of all 15 mice in each treatment group: 0, no hyperpigmentation; 4, maximum hyperpigmentation

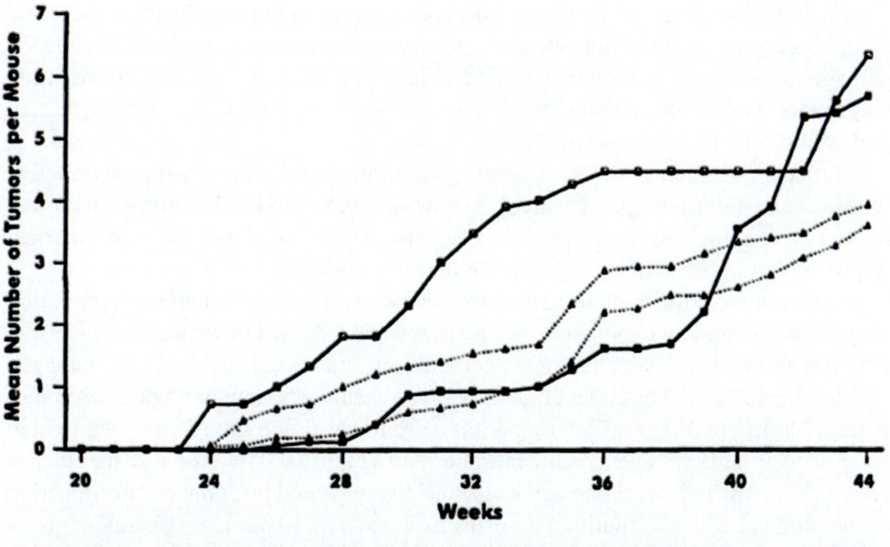

Fig. 4.5 Mean number of tumors ≥2 mm in ultraviolet B (UVB)-irradiated Skh:2 female mice treated with vitamin E and vitamin E esters. Beginning 2 weeks before UV exposure, 15 mice in each treatment group were treated thrice weekly throughout the duration of the experiment with vehicle lotion (*open squares* and *filled squares*), topical d-α-tocopherol lotion (5%) (*filled triangles*), or topical d-α-tocopheryl succinate (5%) (*open triangles*) ≥30 min before UV exposure. In addition, one group (*filled squares*) was fed a diet supplemented with d-α-tocopheryl acetate. The mice were exposed to UVB radiation thrice weekly for 24 weeks, and topical and oral treatments were continued. The number of tumors ≥2 mm on each mouse was counted, and the mean number of tumors per mouse was calculated based on the total number of the mice in each treatment group (Burke et al. 2000)

sunburn and tanning (Burke et al. 2000). Similarly, topical d-α-tocopherol was most effective in decreasing the incidence of UV-induced skin cancer, although all three forms of vitamin E were effective at earlier time points, with topical d-α-tocopheryl succinate almost as effective long term as topical d-α-tocopherol, as seen in Fig. 4.5 (Burke et al. 2000).

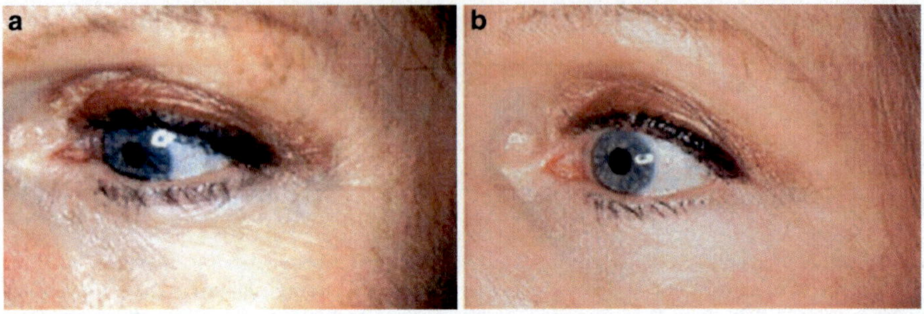

Fig. 4.6 This woman in her early 1940s was photographed (**a**) before treatment and (**b**) after treatment with topical d-α-tocopherol (5%) once daily for 4 months. Note the marked improvement with a decrease in her periorbital wrinkles (**b**)

A new formulation of vitamin E with both dl-α-tocopherol in lecithin and dl-α-tocopheryl ferulate in lecithin showed a measureable decrease in UV-induced DNA damage as measured by reduced d-OHdG (Ichihashi et al. 1999). This formulation was found to inhibit the enzyme tyrosinase and to decrease melanogenesis in human melanoma cells (Ichihashi et al. 1999).

Clinically, vitamin E reverses photoaging dramatically, decreasing unattractive wrinkles and solar lentigos. Figure 4.6 shows a woman in her late forties who was treated with topical d-α-tocopherol (5%) once daily for 4 months. The marked improvement in her skin tone and in the periorbital wrinkles is impressive.

This correction of UV-induced damage was confirmed by histological examination of mice photoaged by exposure to UVB for 6 weeks. After UV exposure, epidermal hypertrophy with thickened stratum corneum, an increased incidence of damaged "sunburn cells" in the basal layer, disruption of dermal collagen with degradation of dermal elastin, and dermal inflammation were noted. Each group was then treated for 8 weeks with vehicle cream, retinoic acid cream (0.05%), or d-α-tocopherol cream (5%). The degree of damage was subjectively graded by "blinded" examination of histological slides of multiple biopsies from each of 10 mice per treatment group by two experienced dermatopathologists. Each parameter—epidermal thickness, hyperkeratosis, collagen disruption, solar elastosis—was assessed separately on a scale of 0 (no damage) to 4 (maximum damage), and a net score was determined, as shown in Fig. 4.7.

Histological improvement in all parameters of photoaging was noted, with a marked decrease in hyperkeratosis and epidermal hypertrophy, repair of damaged dermal collagen and elastin, and clearing of dermal inflammation after treatment with retinoic acid or with d-α-tocopherol. In this limited experiment, topical d-α-tocopherol was shown to be even more effective in reversing photoaging than retinoic acid, the topical medication considered to be the "gold standard" for treatment of photoaging (K.E. Burke, L. Ricotti, E.G. Gross, unpublished observations).

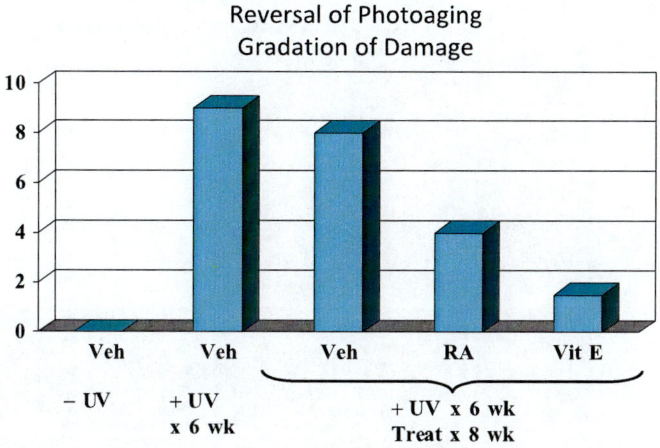

Fig. 4.7 Reversal of photoaging with topical d-α-tocopherol and retinoic acid. The experiment yielding these results is described in detail in the text (K.E. Burke, L. Ricotti, E.G. Gross, unpublished observations)

4.4 Vitamin C with Vitamin E

Vitamins C and E act synergistically in cells to provide antioxidant protection: Vitamin E is located in the cellular membranes, where it quenches free radicals, and vitamin C is plentiful in the neighboring aqueous cytoplasm. With a lower redox potential, vitamin C can reduce the oxidized vitamin E, thereby regenerating vitamin E activity and eliminating the need for nutritional replacement (Chan 1993). In high doses, vitamin C with vitamin E can protect against UV-induced erythema in humans (Eberlein-Konig et al. 1998; Fuchs and Kern 1998), whereas either vitamin alone is ineffective (Fuchs and Kern 1998). Formulating L-ascorbic acid (15%) with α-tocopherol (1%) was found to give fourfold protection against UV-induced erythema in porcine skin, as seen in Fig. 4.8. A decrease in the number of damaged "sunburn cells" was seen histologically as was a decrease in thiamine dimer formation (Lin et al. 2003) compared to twofold protection with either vitamin alone. This protection from UV-induced erythema (Dreher et al. 1998) and tanning (Quevedo et al. 2000) by vitamins C and E was further demonstrated in humans in a formulation containing also melatonin (Dreher et al. 1998). Mixing hydrophilic vitamin C with lipophilic vitamin E has the additional advantage of stabilizing each (Lin et al. 2003). This formulation for topical application is cosmetically attractive and moisturizing.

4.5 Vitamin C with Vitamin E and Ferulic Acid

Ferulic acid is found ubiquitously and at high concentrations in plants (Graf 1992; Rice-Evans et al. 1996; Ou and Kwok 2004), where it crosslinks polysaccharides and proteins during lignin cell wall synthesis (Mathew and Abraham 2004).

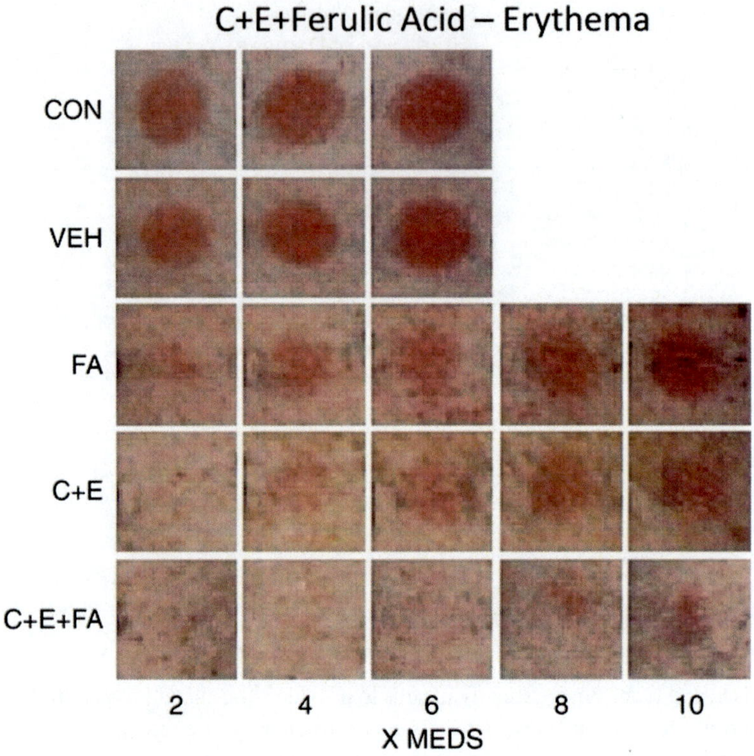

Fig. 4.8 Photoprotection by topical vitamins C and E and ferulic acid. Skin was untreated (**i**) or pretreated with vehicle (**ii**), 0.5% ferulic acid (**iii**), 15% vitamin C + 1% vitamin E (**iv**), or 15% vitamin C + 1% vitamin E + 0.5% ferulic acid (**v**) and irradiated with solar-simulated radiation (2× to 10× the minimum erythema dose). There was visible erythema 1 day later (Photograph by Sheldon R. Pinnell, MD and printed with his permission.)

Ferulic acid is a potent antioxidant, so it further protects membranes from lipid peroxidation and neutralizes alkoxyl and peroxyl radicals. It has also been shown to interact synergistically with ascorbic acid (Trombino et al. 2004). Unlike vitamins C and E, ferulic acid minimally blocks UVB, directly acting as a sunscreen, as seen in Fig. 4.8.

Zielinski and Pinnell (2004) tested the effectiveness of a series of low-molecular-weight antioxidants that are available in chemically pure form. Ferulic acid was found to provide stability of more than 90% for L-ascorbic acid and 100% for α-tocopherol. Addition of ferulic acid (optimally 0.5%) to the formulation of vitamin C (15%) + vitamin E (1%) doubled the photoprotection against solar-simulated irradiation of skin from fourfold to approximately eightfold, as measured by both erythema (with a substantial increase in the MED, as shown in Fig. 4.8) and a decrease in sunburn cell formation (Lin et al. 2005; Murray et al. 2008).

Enhanced photoprotection was further demonstrated immunohistochemically by inhibition of UV-induced formation of thymine dimer mutations and of UV-induced p53, both of which are associated with skin cancer. Evaluation by a real-time polymerase chain reaction demonstrated suppression of UV-induced cytokine mRNA formation (for inflammatory cytokines IL-1α, IL-6, IL-8, and TNFα, as well as for the immunosuppressive cytokine IL-10) (Murray et al. 2008).

The formulation of vitamin C (15%) + vitamin E (1%) + ferulic acid (0.5%) proved to be highly effective in preventing UVB-induced skin cancer in mice (K.E. Burke, X. Zhou, Y. Wang, M. Lebwohl, J. Commisso, K.L. Keen, R.M. Nakamura, H. Wei, manuscript in preparation, 2010). The mice were treated once daily (5 days/week) for 2 weeks prior to exposure to UVB (three times a week for 22 weeks) and throughout the experiment of 35 weeks. One group of 15 mice—vitamin C (15%) + vitamin E (1%) + ferulic aid (0.5%)—were treated with vehicle serum and the other with antioxidant serum. Amazingly, only one tumor was seen in the antioxidant-treated group, whereas the vehicle-treated group had 195 tumors. To compare these figures with data from a similar experiment (shown in Fig. 4.5) (Burke et al. 2000), after 40 weeks of observation the group of 15 vehicle-treated mice had 67 tumors cumulatively and the 15 d-α-tocopherol-treated mice had 36 tumors.

The data presented above demonstrates the remarkable protection possible with these topical antioxidants, vitamins C and E. Other new formulations with vitamins C and E in microsomes or in microemulsions are equally impressive in protecting against UV-induced sunburn, tanning, and skin cancer. Many other antioxidants and their combinations are being actively investigated.

Take-Home Messages

- Because sunscreen alone cannot provide enough protection from UV-induced and other environmental free-radical damage to the skin, additional protection of the skin from both UVA and UVB and from the oxidative damage of pollutants is essential to maintain healthy, young skin and to prevent precancers and cancers.
- When formulated correctly, the topical antioxidants vitamins C and E not only protect against UV-induced and other free-radical damage, they can also reverse previous photoaging.
- The most protective formulation available today is a serum that contains vitamin C (L-ascorbic acid 15%) with vitamin E (α-tocopherol 1%) and ferulic acid (0.5%). Other formulations of vitamins C and E together in microsomes or in microemulsions show similar efficacy. Many other antioxidants and their combinations are currently being investigated and show great potential.
- Because topical antioxidants are absorbed by the skin to become a reservoir of protection that is not lost by washing or by perspiration, one application per day is sufficient to enhance the protection by sunscreen against acute sunburn and tanning as well as against chronic photoaging, precancers and cancers.

References

Alster T, West TB. Effect of vitamin C on postoperative CO_2- laser resurfacing erythema. Dermatol Surg. 1998;24:331–4.

Azzi A, Breyer I, Feher M, et al. Specific cellular responses to alpha-tocopherol. J Nutr. 2000;131(suppl):369–75.

Beijersbergen van Hanegouwne GM, Junginger HE, de Vries H. Hydrolysis of RRR-alpha-tocopheryl acetate (vitamin E acetate) in the skin and its UV protecting activity (an in vivo study with the rat). J Photochem Photobiol B. 1995;29:45–51.

Bergfeld W, Pinnell S. Topical vitamin C. Dialogues in dermatology. Am Acad Dermatol. 1996;38:1.

Berton TR, Conti CJ, Mitchell DL, et al. The effect of vitamin E acetate on ultraviolet-induced mouse skin carcinogenesis. Mol Carcinog. 1998;23:175–84.

Bissett DL, Chatterjee R, Hannon DP. Photoprotective effect of superoxide-scavenging antioxidants against ultraviolet radiation-induced chronic skin damage in the hairless mouse. Photodermatol Photoimmunol Photomed. 1990;7:56–62.

Bissett D, Chatterjee R, Hannon D. Protective effect of a topically applied antioxidant plus an antiinflammatory agent against ultraviolet radiation-induced chronic skin damage in the hairless mouse. J Soc Cosmet Chem. 1992a;43:85–92.

Bissett D, Chatterjee R, Hannon D. Protective effect of a topically applied antioxidant plus an anti-inflammatory agent against ultraviolet radiation-induced chronic skin damage in the hairless mouse. J Soc Cosmet Chem. 1992b;43:65–92.

Burke KE, Wei H. Synergistic damage by UVA radiation and pollutants. Toxicol Ind Health. 2009;25:219–24.

Burke DE, Clive J, Combs GF Jr, et al. The effects of topical and oral vitamin E on pigmentation and skin cancer induced by ultraviolet irradiation in Skh:2 hairless mice. Nutr Cancer. 2000;38:87–97. Published by Taylor and Francis, Ltd. www.informaworld.com.

Burton GW, Traber MG, Acuff RV, et al. Human plasma and tissue -tocopherol concentrations in response to supplementation with deuterated natural and synthetic vitamin E. Am J Clin Nutr. 1998;67:669–84.

Carcamo JM, Pedraza A, Borquez-Ojeda O, Golde DS. Vitamin C suppresses TNF alpha-induced NF kappa B activation by inhibiting I Kappa B alpha phosphorylation. Biochemistry. 2002;41: 12995–30002.

Chan AC. Partners in defense, vitamin E and vitamin C. Can J Physiol Pharmacol. 1993;71:725–31.

Darr D, Combs S, Dunsten S, et al. Topical vitamin C protects porcine skin from ultraviolet radiation-induced damage. Br J Dermatol. 1992;127:247–53.

Davidson JM, Luvalle PA, Zoia O, et al. Ascorbate differentially regulates elastin and collagen biosynthesis in vascular smooth muscle cells and skin fibroblasts by pretranslational mechanisms. J Biol Chem. 1997;272:345–52.

Dreher F, Gabard B, Schwindt DA, Malbach HI. Topical melatonin in combination with vitamins E and C protects skin from ultraviolet-induced erythema: a human study in vivo. Br J Dermatol. 1998;139:332–9.

Eberlein-Konig B, Placzek M, Przybilla B. Protective effect against sunburn of combined systemic ascorbic acid (vitamin C) and d-α-tocopherol (vitamin E). J Am Acad Dermatol. 1998;38:45–8.

Fuchs J, Kern H. Modulation of UV-light-induced skin inflammation by d-α-tocopherol and L-ascorbic acid: a clinical study using solar simulated radiation. Free Radic Biol Med. 1998;25:1006–12.

Gensler H, Magdaleno M. Topical vitamin E inhibition of immunosuppression and tumorigenesis induced by ultraviolet irradiation. Nutr Cancer. 1991;15:97–106.

Gensler HL, Aickin M, Peng YM, Xu M. Importance of the form of topical vitamin E for prevention of photocarcinogenesis. Nutr Cancer. 1996;26:183–91.

Gerrish K, Gensler H. Prevention of photocarcinogenesis by dietary vitamin E. Nutr Cancer. 1993;19:125–33.

Graf E. Antioxidant potential of ferulic acid. Free Radic Biol Med. 1992;13:435–48.

Hart M. Vitamin E: a contact sensitizer. Schoch Lett. 1990;40:48.
Haywood R, Wardman P, Sanders R, Linge C. Sunscreens inadequately protect against ultraviolet A-induced free radicals in skin: implications for skin aging and melanoma? J Invest Dermatol. 2003;121:862–8.
Humbert PG, Haftek M, Creidi P, et al. Topical ascorbic acid in photoaged skin. Clinical topographical and ultrastructural evaluation: double-blind study vs. placebo. Exp Dermatol. 2003;12:237–44.
Ichihashi M, Funasaka Y, Ohashi A, et al. The inhibitory effect of DL-alpha-tocopheryl ferulate in lecithin on melanogenesis. Anticancer Res. 1999;19:3769–74.
Kameyama K, Sakai C, Kondoh S, et al. Inhibitory effect of magnesium L-ascorbyl-2-phosphate (VC-PMG) on melanogenesis in vitro and in vivo. J Am Acad Dermatol. 1996;34:29–33.
Kivirikko KI, Myllyla R. Post-translational processing of procollagens. Ann NY Acad Sci. 1996;11:250–3.
Lin J, Selim A, Shea C, et al. UV photoprotection by combination topical antioxidants vitamin C and E. J Am Acad Dermatol. 2003;48:866–74.
Lin FH, Lin JY, Gupta RD, et al. Ferulic acid stabilizes a solution of vitamins C and E and doubles its photoprotection of skin. J Invest Dermatol. 2005;125:826–32.
Maeda K, Fukuda M. Arbutin: mechanism of its depigmenting action in human melanocyte culture. J Pharmacol Exp Ther. 1996;276:765–9.
Marenus K, Muizzuddin N, Kasman K, et al. The use of antioxidants in providing protection from chronic suberythemal UV-B exposure. 16th IFSCC Conference, Oct 1990;1:24–34.
Mathew S, Abraham TE. Ferulic acid: an antioxidant found naturally in plant cell walls and feruloyl esterases involved in its release and their applications. Crit Rev Biotechnol. 2004;24:59–83.
Mayer P, Pittermann W, Wallat S. The effects of vitamin E on the skin. Cosmet Toiletries. 1993;108:99–109.
Murray JC, Burch JA, Streilein RD, et al. A topical antioxidant solution containing vitamins C and E stabilized by ferulic acid provides protection for human skin against damage caused by ultraviolet irradiation. J Am Acad Dermatol. 2008;59:418–25.
Ou S, Kwok K-C. Ferulic acid: pharmaceutical functions preparation and applications in foods. J Sci Food Agric. 2004;84:1261–9.
Phillips CL, Combs SB, Pinnell SR. Effects of ascorbic acid on proliferation and collagen synthesis in relation to donor age of human dermal fibroblasts. J Invest Dermatol. 1994;103:228–32.
Pinnell SR, Yang HS, Omar M, et al. Topical L-ascorbic acid: percutaneous absorption studies. Dermatol Surg. 2001;27:137–42.
Podda M, Weber C, Traber MG, Packer L. Simultaneous determination of tissue tocopherols, tocotrienols, ubiquinols, and ubiquinones. J Lipid Res. 1996;37:893–901.
Potokar M, Holtmann W, Werner-Busse A. Effectiveness of vitamin E protecting against UV light – comparative testing of the natural tocopherols on the skin of the hairless mouse. Fat Sci Technol. 1990;92:406–10.
Puling L. How to live longer and feel better. New York: WH Freeman; 1987.
Quevedo WC, Holstein TJ, Dyckman J, McDonald CI. The responses of the human epidermal melanocyte system to chronic erythemal doses of UVR in skin protected by topical applications of a combination of vitamins C and E. Pigment Cell Res. 2000;13:190–2.
Record IR, Dreosti IE, Konstantinopoulos M, Buckley RA. The Influence of topical and systemic vitamin E on ultraviolet light-induced skin damage in hairless mice. Nutr Cancer. 1991;16: 219–25.
Rice-Evans CA, Miller NJ, Paganga G. Structure-antioxidant activity relationships of flavonoids and phenolic acids. Free Radic Biol Med. 1996;20:933–56.
Roshchupkin D, Pistsov MY, Potapenko AY. Inhibition of ultraviolet light-induced erythema by antioxidants. Arch Dermatol Res. 1979;266:91–4.
Saperstein H, Rapaport M, Rietschel RL. Topical vitamin E as a cause of erythema multiforme-like eruption. Arch Dermatol. 1984;120:906–8.
Savini I, Catni V, Rossi A, Duranti G, Melino G, Avigliano L. Characterization of keratinocyte differentiation induced by ascorbic acid: protein kinase C involvement and vitamin C homeostasis. J Invest Dermatol. 2002;118:372–9.

Shindo Y, Wit E, Han D, Packer L. Dose response effects of acute ultraviolet irradiation on antioxidants and molecular markers of oxidation in murine epidermis and dermis. J Invest Dermatol. 1994;23:470–5.

The Skin Cancer Foundation – Guidelines. http://www.skincancer.org/Guidelines/ (2010). Accessed July 2010.

Thiele JJ. Oxidative targets in the stratum corneum: a new basis for antioxidative strategies. Skin Pharmacol Appl Skin Physiol. 2001;14:87–91.

Thiele JJ, Traber MG, Tsange KG, et al. In vivo exposure to ozone depletes vitamins C and E and induces lipid peroxidation in epidermal layers of murine skin. Free Radic Biol Med. 1997;23:85–91.

Trevithick J, Xiong H, Lee S, et al. Topical tocopherol acetate reduces post-UVB, sunburn-associated erythema, edema and skin sensitivity in hairless mice. Arch Biochem Biophys. 1992;296:575–82.

Trombino S, Serini S, Di Nicuolo F, et al. Antioxidant effect of ferulic acid in isolated membranes and intact cells: synergistic interactions with alpha-tocopherol, beta-carotene, and ascorbic acid. J Agric Food Chem. 2004;52:2411–20.

Uchida Y, Behne M, Quiec D, et al. Vitamin C stimulates sphingolipid production and markers of barrier formation in submerged human keratinocyte cultures. J Invest Dermatol. 2001;117:1307–13.

Wang Y, Saladi R, Wei H. Synergistic carcinogenesis of chemical carcinogens and long wavelength UVA radiation. Trends Photochem Photobiol. 2003;10:31–45.

Wulf HC, Stender IM, Lock-Andersen J. Sunscreens used at the beach do not protect against erythema: a new definition of SPF is proposed. Photodermatol Photoimmunol Photomed. 1997;13:129–32.

Zielinski JE, Pinnell SR. Stabilized ascorbic acid compositions and methods thereof. US Patent 34405.2, 2004.

Chapter 5
Carotenoids and Skin

Sagar K. Thakkar, Angus M. Moodycliffe, and Myriam Richelle

Core Messages

- Dietary carotenoid bioavailability is rather low in comparison to other macronutrients. However, they are absorbed and distributed to hepatic and some extrahepatic tissues, including skin.
- Co-consumption of dietary fat is essential for carotenoid bioavailability. In contrast, the presence of some other dietary constituents, such as fiber, may reduce carotenoid bioavailability.
- Fruits and vegetables contain bioactive agents, including carotenoids, that protect not only plants but also humans against solar ultraviolet (UV) radiation damage. Only recently has the role of dietary bioactive agents in the photoprotection of skin started to be investigated and appreciated.
- Systemic photoprotection via a dietary supply of carotenoids may contribute significantly to skin health and complement the use of sunscreens in protecting the skin against the damaging effects of solar UV exposure.
- However, the knowledge acquired so far about the role of dietary carotenoids in photoprotection is still in its infancy and needs to be further investigated.

5.1 Introduction

In mammals, the skin is the largest organ system. Collectively and individually, all layers of the skin work toward the primary goal of protecting internal tissues and vital organs from day-to-day environmental challenges. Because of the presence of

M. Richelle (✉)
Nestlé Research Center, Vers-Chez-Les-Blanc,
PO Box 44, CH-1000, Lausanne, 26, Switzerland
e-mail: myriam.richelle@rdls.nestle.com

A. Pappas (ed.), *Nutrition and Skin: Lessons for Anti-Aging, Beauty and Healthy Skin*,
DOI 10.1007/978-1-4419-7967-4_5, © Springer Science+Business Media, LLC 2011

nerve endings, the skin also plays a role in the detection of sensations such as heat, cold, touch, pressure, and injury. Skin also plays an important role in the regulation of body temperature and controls water and mineral loss. The synthesis and storage of lipophilic molecules such as vitamin D are essential functions of the skin. On a daily basis, skin is exposed to various chemical and physical agents that may harm its integrity and lead to a variety of skin pathologies including skin cancer. Visible signs of premature skin aging are attributed to the repeated exposure of the skin to environmental insults, including solar ultraviolet (UV) radiation, leading to the generation of oxidative free-radical molecules, which can damage cellular lipids, proteins, and DNA, —thereby influencing cell survival or death (Fisher et al. 2002).

For centuries, humans have knowingly or unknowingly been protecting their skin against the detrimental effects of solar UV exposure, either by wearing protective clothing, avoiding sun exposure, topically applying herbal pastes and oils, and ingesting fruits and vegetables that contain bioactive compounds that protect not only plants but also their consumer against UV damage. Despite this, only recently has the role of dietary bioactive agents in photoprotection of skin started to gain appreciation based on scientific investigation. Mounting evidence from nutritional research suggests that many dietary nutrients accumulate in the skin and may provide endogenous protection. Deficiencies of vitamin A, vitamin C, riboflavin, niacin, pyridoxine, vitamin E, zinc, selenium, and certain fatty acids have been shown to cause skin anomalies. The roles of some of these nutrients are well characterized; for example, ascorbate is an essential cofactor in collagen synthesis, and vitamin A is required for gene transcription. In contrast, a balanced nutritional diet and a diet rich in nutrients such as polyphenols and omega-3 fatty acids (e.g., docosahexaenoic acid [DHA] and eicosapentaenoic acid [EPA]) have been proposed to maintain and improve overall skin health. On the other hand, we are just starting to understand how carotenoids and their metabolites lend their chemical structure to help protect skin.

Epidemiological studies from regions with high consumption of carotenoid-rich fruits and vegetables have shown lower incidences of skin pathology. In vitro and in vivo evidence also indicates the role of endogenous carotenoids in systemic protection and maintenance of skin health (Garmyn et al. 1995; Sies and Stahl 2004). Hence, for the benefit of the readers, we put forward the existing evidence of protective roles of carotenoids in skin and raise issues where we think there is a potential to obtain concrete scientific evidence.

5.2 Dietary Carotenoids

Carotenoids comprise a family of highly lipophilic pigments that are synthesized by all photosynthetic organisms and some nonphotosynthetic micro-organisms, but not animals. Most carotenoids have a polyisoprenoid structure with a long chain of conjugated double bonds and nearly bilateral symmetry around the central double bond (Fig. 5.1). Various classes are derived by modification of the basic structure.

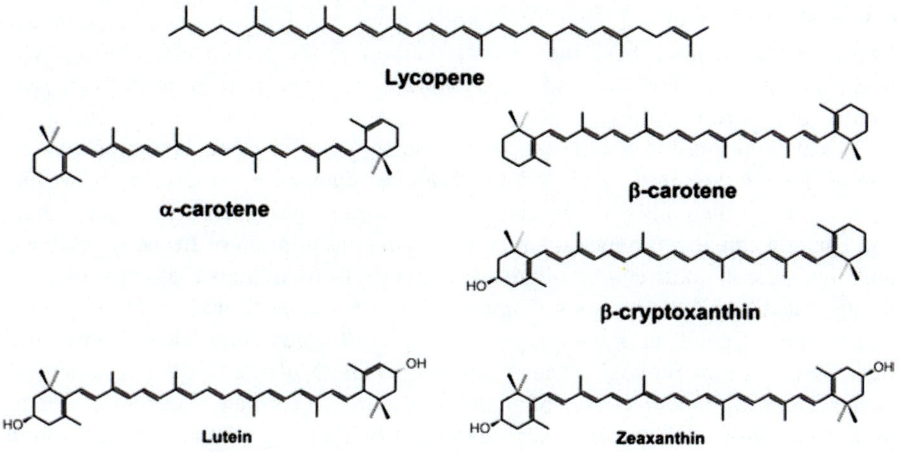

Fig. 5.1 Structures of the most abundant dietary carotenoids

For example, cyclization of end groups and the introduction of functional groups is widespread and results in characteristic colors and unique biological properties. The pattern of conjugated double bonds in the polyene backbone of carotenoids is responsible for their light-absorbing properties and influences the antioxidant activity of specific carotenoids. Several *cis/trans* configurations at double bonds are thermodynamically feasible for a given carotenoid (Britton 1995). Generally the *trans* isomer is the predominant form in nature, but exposure to light, heat, and acid during the processing of plant foods after harvest and cooking may induce isomerization (Rodriguez-Amaya 2003).

Carotenoids are divided into two major classes: carotenes and xanthophylls. Carotenes, also known as hydrocarbon carotenoids, contain only hydrogen and carbon atoms. The most common dietary carotenes include the acyclic lycopene and its biosynthetic downstream cyclic products β-carotene and α-carotene. Xanthophylls are also known as oxycarotenoids due to the presence of at least one oxygen molecule in the structure. Xanthophylls are more hydrophilic than carotenes. Lutein, zeaxanthin, and β-cryptoxanthin represent the most common xanthophylls found in the human diet. In many fruits and vegetables, xanthophylls may be esterified with one or two long-chain fatty acids depending on the number of esterification sites available. These species are more hydrophobic than their nonesterified analogues owing to the presence of fatty acids.

Because de novo synthesis of carotenoids is not possible, humans have to depend on dietary sources to obtain them. Carotenoids appear in shades of yellow and orange to red, and they are commonly found in fruits and vegetables of those colors. Orange fruits and vegetables (e.g., carrot, mango, orange-flesh sweet potato, apricots, squash) contain β-carotene, α-carotene, and β-cryptoxanthin, which are the three most common provitamin A carotenoids. Populations solely depending on fruits and vegetables as a source of their energy obtain their vitamin A from consumption of such foods. On the other hand, green vegetables (e.g., spinach, kale, collard greens), which contain carotenoids and are excellent sources of β-carotene, lutein, and zeaxanthin, have their colors masked by the presence of chlorophyll. Red fruits and vegetables (e.g., tomato, watermelon, guava, pink grapefruit) are rich sources of lycopene.

The main purpose of the existence of these pigments in plants is to preserve the photosynthetic complex by protecting it against damage by solar UV. However, synthesis of carotenoids is not limited to plants; certain photosynthetic algae, fungi, and bacteria can also produce them. Apart from consumption of fruits, vegetables, and algae, carotenoids can be obtained indirectly from animals that consume and accumulate them in their tissues. Commonly consumed sea animals, such as salmon and shrimp, depend on aquatic algae rich in carotenoids consequently enriching their tissue with carotenoids. These carotenoids are available to the consumers of such seafood. Another example of an indirect source of carotenoids is enriched egg yolk. Depending on the type of diet fed to chicken, their egg yolks are either orange or yellow, which are rich in β-carotene and lutein, respectively. Over the last few decades, more than 600 carotenoids found in nature have been well characterized. However, humans consume only about 50 carotenoids from various sources in their diet, and approximately half of them are found in human plasma (Khachik et al. 1992). It is evident that the bioavailability of carotenoids in plasma must precede its distribution to hepatic and extrahepatic tissues including skin.

5.3 Carotenoids Bioavailability and Biodistribution to Skin

5.3.1 Carotenoids Bioavailability

Because carotenoids are lipophilic molecules, their patterns of absorption and transport are similar to those for other dietary lipids. Bioavailability of carotenoids is defined as a transfer of dietary carotenoids and their metabolites to the lymphatic or portal circulation for distribution to hepatic and extrahepatic tissues for biological functioning, metabolism, or storage. Uptake by intestinal mucosal cells in itself cannot be categorized as bioavailability because mucosal cells may be sloughed off into the lumen before carotenoids or their metabolites can cross the basolateral surface. Therefore, studies of carotenoid absorption are often complex because individuals consume carotenoids as components of meals. Food matrices and the presence of competing molecules in the diet affect the efficiency of the transfer of carotenoids to intestinal absorptive cells, thereby making reliable prediction of absorption difficult. As outlined in Fig. 5.2, absorption of carotenoids from a meal requires several processes, including (1) release of carotenoids from a food matrix, (2) incorporation into lipid droplets, (3) transfer to mixed bile salt micelles in the lumen of the small intestine, (4) uptake of carotenoid molecules across the apical surface of intestinal mucosal cells from bile salt micelles, (5) incorporation of carotenoids and their metabolites into chylomicrons, and (6) efflux of chylomicrons across the basolateral membrane into the lymphatic circulation.

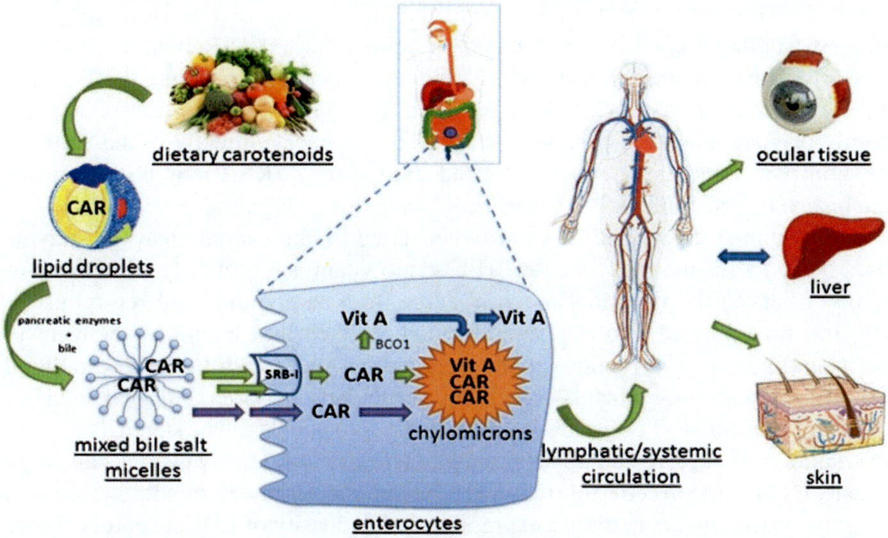

Fig. 5.2 Carotenoid bioavailability and biodistribution to hepatic and extrahepatic tissues. **CAR* carotenoids, *Vit A* Vitamin A, *BCO1* β-carotene 15,15′-monooxygenase 1; *SRB-I* scavenger receptor type B class I

Release of carotenoids from the matrix in which it is embedded begins by disruption during processing and cooking of the food. Reducing the particle size by purée-ing or chopping vegetables results in relatively higher in vitro bioaccessibility and higher potential for improved bioavailability than that in whole or sliced raw vegetables (Ryan et al. 2008; Thakkar et al. 2009). Further disruption of the matrix occurs during mastication and mixing of food with salivary juices and enzymes. Hydrochloric acid, proteolytic enzymes, gastric lipases, and peristaltic movements of the upper gastrointestinal tract further disrupt the food matrix for release of carotenoids. Release from the food matrix generally results in partitioning of carotenoids into lipid droplets available in the gastric lumen. Localization of carotenoids in lipid droplets is dependent on their structure, with polar carotenoids distributed on the surface and the more hydrophobic carotenoids such as carotenes localized in the core (Borel et al. 1996).

The presence of dietary lipid stimulates bile flow from the gallbladder, which facilitates emulsification of lipid droplets and other lipophilic molecules in the small intestine. Pancreatic lipases hydrolyze the lipid droplets to much smaller particles. This hydrolytic activity leads to conversion of substrates such as triglycerides, phospholipids, cholesterol esters, retinyl esters, and carotenoid esters to free fatty acids monoacylglycerides, lysophospholipids, free cholesterol, retinol, and free carotenoids, respectively. Incorporation of carotenoids in mixed bile salt micelles is obligatory for transfer of carotenoids across the unstirred water layer to the apical surface of absorptive cells. Owing to aforementioned rationales, the presence of dietary lipid in a carotenoid-enriched meal is of utmost importance for its bioavailability. Carotenoid uptake into intestinal epithelial cells has been assumed to occur passively. However, recent studies from the laboratories of Harrison and Borel suggest facilitated uptake of carotenoids (During et al. 2005; Reboul et al. 2005). Several apical membrane transporters, including Nieman-Pick C1-Like 1 (NPC1L1), scavenger receptor type B class I (SRB-I), and ATP binding cassette transporter sub-family A (ABCA1), have been hypothesized to participate directly or indirectly in the transport. Among the above-indicated transporters, SRB-I may be of primary interest, yet its role in uptake of carotenoids remains ambiguous.

Provitamin A carotenoids are partly converted by the central cleavage enzyme β-carotene 15,15′-monooxygenase 1 (BCO1) to vitamin A primarily in the form of retinyl esters in the intestinal mucosal cells. Both carotenoids and retinyl esters are then incorporated into chylomicrons and secreted into lymph for delivery to peripheral tissues. Chylomicrons are modified by lipoprotein lipase in the blood to form chylomicron remnants containing carotenoids that are rapidly taken up by the liver (Yonekura and Nagao 2007). In the liver, carotenoids can be stored, converted to vitamin A, or resecreted into the circulation in very low density (VLDL) and high density (HDL) lipoproteins for transport to peripheral tissues. Carotenoids appear to selectively accumulate in tissues expressing a high density of LDL receptors. These tissues include the adrenals, testes, liver, adipose tissue, kidney, and skin (Meinke et al. 2010; Gerster 1997). An exception may be the selective delivery of lutein and zeaxanthin to the macula region of the retina by HDL (Connor et al. 2007).

5.3.2 *Factors Influencing Carotenoid Bioavailability*

The bioavailability of carotenoids is dependent on physiochemical properties of the carotenoid (e.g., crystalline versus solubilized in plant organelles and free versus protein-bound), food matrix (e.g., located in chloroplast versus chromoplast; root versus leaf versus seed as food), type of processing of raw food (e.g., sun-dried, fermented, boiled, fried), presence or absence of compounds that promote or inhibit their absorption (e.g., lipids, fiber, other carotenoids), pathophysiological status of the gut (e.g., malabsorption due to parasites, pancreatic insufficiency, cholestasis), and nutritional status (e.g., deficient or adequate vitamin A) of the individual. These factors are further discussed below.

5.3.2.1 Physicochemical Properties of Carotenoids

Carotenoids can accumulate in plant foods in either crystalline or solubilized form. Carotenoid solubilized in oil droplets are transferred to the micelle fraction more efficiently than those in crystalline form (Rich et al. 2003a, b). Carotenoid speciation also appears to affect bioavailability. Oxycarotenoids are more hydrophilic owing to the presence of one or more polar functional groups and are located on the surface of oil droplets. Hydrocarbon carotenoids are more hydrophobic in comparison and are embedded in the core. This decreases the efficiency of transfer of the hydrocarbon carotenoid to the micelle fraction as compared to oxycarotenoids (Borel et al. 1996; van het Hof et al. 1999). Xanthophylls in certain foods can also exist as their fatty acid esters. These esters are extremely hydrophobic in nature and require hydrolysis to free carotenoids for efficient partitioning into the micelle fraction and uptake by enterocytes. Small intestinal simulations have demonstrated that hydrolytic cleavage of ester bond by carboxyl ester lipase, a broad-specificity pancreatic enzyme, precedes the preferential uptake of the nonesterified form of carotenoids by small intestinal cells (Chitchumroonchokchai and Failla 2006).

5.3.2.2 Food Matrix and Processing

Carotenoids may be bound to proteins in the food matrix. Food processing alters the matrix by disrupting the cell wall and membrane-bound organelles, loosening the linkages between carotenoids and proteins or fiber, partially dissolving the crystalline carotenoids in oil, and increasing its surface area. These changes increase access of digestive enzymes and bile salts to promote the partitioning of carotenoids into micelles to enhance their bioavailability (Castenmiller et al. 1999; Edwards et al. 2002; Livny et al. 2003). Stahl and Sies demonstrated that lycopene concentrations in human serum increased only when tomatoes were subjected to 1 h of boiling in the presence of oil (Stahl and Sies 1992). Processing of plant foods also induces isomerization of carotenoids, thus increasing the levels of *cis* isomers. For example, baking is associated with isomerization and degradation of all-*trans* β-carotene in sweet potatoes (Chandler and Schwartz 1988). The percent bioavailability of *cis* analogues have been reported to be higher than their *trans* counterparts, although it should be noted that quantitatively the isomeric profile in plasma is still a reflection of the profile in food (Stahl and Sies 1992; Stahl et al. 1992). Although *cis* isomers of β-carotene are also precursors of vitamin A, isomerization of a provitamin A carotenoid reduces the retinol activity equivalence (RAE) as compared to their *trans* analogues (Deming et al. 2002; National Academy of S and M Institute 2001).

5.3.2.3 Interactions with Other Dietary Components

Consumption of carotenoids in our diet is usually associated with other macrontrients and micronutrients in the diet, which may act as promoters or inhibitors of carotenoid bioavailability. The association of carotenoids with dietary proteins has been shown to decrease their absorption in ferrets (Sundaresan et al. 2005). Likewise, dietary fibers such as citrus pectin and wheat bran are likely to bind to bile salts and decrease the formation of micelles and ultimately the absorption of carotenoids (Zanutto et al. 2002). In vivo and in vitro studies have demonstrated that dietary fat promotes carotenoid absorption. For example, Unlu and associates demonstrated that carotenoid absorption from a salad meal was enhanced by addition of avocado or avocado oil (Unlu et al. 2005). Several mechanisms have been proposed to be responsible for this effect. Dietary fat may facilitate the release of carotenoids from the food matrix by serving as a sink for hydrophobic compounds, by stimulating the secretion of bile and pancreatic enzymes, and by enhancing micelle formation in the small intestine. Moreover, the synthesis and secretion of chylomicrons by enterocytes is increased by dietary fat (Ribaya-Mercado 2002). Therefore, co-consumption of dietary lipid has been proposed to enhance the bioavailability of carotenoids.

Some, but not all, investigators have reported negative interactions between dietary carotenoids. High doses of pure β-carotene and lutein have been reported to antagonize the absorption of one another (Kostic et al. 1995; van den Berg and van Vliet 1998). The mechanisms for preabsorptive interactions among carotenoids are not well understood. Proposed interactions have been suggested to occur during their incorporation into bile salt micelles, uptake by intestinal epithelial cells, and incorporation into chylomicrons within enterocytes (van den Berg 1999).

5.3.2.4 Health Status of the Host

The digestive health status, and indeed the overall health status of an individual, also influences the absorption of carotenoids. Increased gastric pH appears to decrease β-carotene absorption (Tang et al. 1996). Gastrointestinal conditions that cause fat malabsorption may also lead to decreased carotenoid absorption. Conditions such as cholestasis, biliary cirrhosis, and pancreatic insufficiency may cause a reduction in digestive enzymes and bile release, ultimately affecting overall digestion of carotenoid-rich food and partitioning of lipophilic compounds into micelles (Olson 1999). Parasitemia in general has been negatively associated with plasma concentrations of carotenoids (Metzger et al. 2001). A study with malaria-infested children in Uganda demonstrated that therapeutic reductions in parasitemia resulted in increased concentrations of plasma carotenoids. Also, the presence of parasites in the intestine impaired carotenoid absorption or utilization. Jalal and his co-workers reported enhanced absorption after deworming children infected with *Ascaris* (Jalal et al. 1998).

5.3.3 Carotenoids Biodistribution to Skin

Once carotenoids clear the intestinal epithelial barrier and reach the systemic circulation via lymphatics, deposition occurs in hepatic and extrahepatic tissues including skin. Carotenoids stored in hepatic tissues can also be mobilized for distribution to extrahepatic tissues in times of need. Alaluf and coworkers (Alaluf et al. 2002) have demonstrated that the yellow component of skin color, quantified by the tristimulus chromameter, may be closely associated with carotenoid levels of the skin at various sites of the body (i.e., the back, forehead, inner forearm, and palm of the hand (Bayerl et al. 2003). Skin derived from subjects undergoing abdominoplasty suggests that apolar carotenoids (e.g., β-carotene, phytoene, lycopene) and their isomers constitute approximately 70% of the total pool; xanthophylls such as lutein and zeaxanthin are present at lower concentrations (Hata et al. 2000). However, it should be noted that dietary intake based on the availability of carotenoid-enriched foods and cultural traditions in different regions of the world may influence this profile.

Although stereoisomers of carotenoids mainly exist as the all-*trans* configuration in nature, *cis* isomers are also present in skin (Stahl et al. 1992). A high concentration of *cis* isomers in skin may be due to preferential uptake of *cis* isomers from the circulation, isomerization in biological systems after absorption, or both. Recent data from our laboratory suggests that isomerization begins in small-intestinal epithelial cells and may even occur in other tissues (Richelle et al. 2010). Xanthophylls esters (i.e., lutein, zeaxanthin) are also present in skin as mono- or di-fatty acid esters in picomolar ranges; they may be formed by postprandial reesterification of xanthophylls (Wingerath et al. 1998). Carotenoids exhibit a concentration gradient in the layers of skin with a higher amount in the dermis (inner layer) and lower levels in the stratum corneum (outer layer). However, it is unclear whether stratum corneum has higher utilization or lower deposition when compared to the deeper layers. Regional variations may also be observed in the skin carotenoid level; indeed, higher concentrations of total carotenoids are measured in skin of the forehead, palm of the hand, and dorsal skin, whereas lower levels are present in the arm and the back of the hand (Bayerl et al. 2003; Stahl et al. 1998).

Occasionally, consumption of diets or supplementation containing more than 30 mg carotenoids per day for more than 4 weeks may result in yellowish discoloration of the skin, called carotenodermia. This condition is reversible, with the usual skin color restored upon cessation of the responsible supplementation or of a carotenoid-enriched diet (Bruch-Gerharz et al. 2001; Micozzi et al. 1988; Dimitrov et al. 1988).

5.4 Techniques to Analyze Carotenoids in Skin

To date, the accepted methodology for measuring carotenoids in living human skin tissue involves performing skin biopsies, the specimens from which are then analyzed by high performance liquid chromatography (HPLC) for the presence of carotenoids. As this methodology is highly invasive, the possibility of multiple biopsies or the willingness of volunteers to participate may become difficult. Furthermore, among those who would participate, a high dropout rate may be observed. Despite its invasive nature, chemical analyses by HPLC for carotenoids in skin biopsies from cadavers, abdominoplasty subjects, and a few healthy subjects have been reported (Hata et al. 2000; Vahlquist et al. 1982). Even though limited data from human trials are available, comparison of the data from different trials is a challenging task as different methods were used to obtain skin biopsies, including blister, scrape, shave, and punch biopsies with possible contamination from subcutaneous fat. It is also noteworthy that the complexity of the matrix of the skin biopsies makes extraction of carotenoids for HPLC analysis a challenging task.

More recently, researchers have started to focus on the use of Raman spectroscopy for noninvasively assessing the concentration of carotenoids in skin (Hammond and Wooten 2005; Ermakov et al. 2004, 2005; Hesterberg et al. 2009; Darvin et al. 2009). Although many claims of precision, accuracy, specificity, and sensitivity have been made about the Raman technique, the claims have been challenged multiple times. In this regard, there are concerns that other chromophores in the skin interfere with carotenoid measurements, the distribution of different carotenoids is not uniform across the layers of the skin, and the morphology of the stratum corneum varies among subjects in a cohort.

Another major concern about the use of Raman spectroscopy is its lack of validity with in vivo data. Recently, two laboratories reported a good correlation between the amount of carotenoids in plasma compared to that measured by Raman spectroscopy in skin (Zidichouski et al. 2009; Meinke et al. 2010). However, we believe that the comparison of skin and plasma carotenoids could be coined as a partial validation, as it does not involve chemical analyses of carotenoids in skin tissues and surfaces that were measured by Raman spectroscopy. Nevertheless, this noninvasive method has been used in multiple studies; and pending future validation, it has the potential to be included in routine analyses of carotenoids of the skin.

5.5 Dietary Carotenoids Protect Skin Against Some of the Damaging Effects of UV Exposure

5.5.1 Damaging Effects of UV on Skin

Ultraviolet radiation present in sunlight represents one of the most important environmental hazardous physical agents that the skin encounters on a daily basis throughout a person's lifetime. Depending on the amount and form of the UV radiation and the skin type of the individual exposed, UV irradiation may cause tissue injury and cutaneous inflammation signifying sunburn (Clydesdale et al. 2001; Cavallo and DeLeo 1986), immune suppression (Ullrich 2005; Moodycliffe et al. 2000), premature aging of the skin called photoaging (Helfrich et al. 2008), and skin cancer (Melnikova and Ananthaswamy 2005; Matsumura et al. 2004). Chronic exposure to solar UV is considered the major etiological factor for the development of nonmelanoma skin cancer, which occurs primarily on sun-exposed areas of the body.

Sunburn is a term applied to the marked erythema and pain that commonly follows sun overexposure. A sunburn is delayed UV-induced erythema caused by an increase in blood flow to the affected skin that begins about 4 h after exposure and peaks at 8–24 h (Andersen et al. 1991; Ramsay and Challoner 1976). The underlying cause of sunburn is direct and indirect damage to specific cellular targets from photochemical reactions and the generation of reactive oxygen species (Hruza and Pentland 1993). Damage to DNA, the activation of several inflammatory pathways, and the release of inflammatory mediators by keratinocytes are thought to trigger this reaction, ultimately leading to vasodilation and edema (Clydesdale et al. 2001; Cavallo and DeLeo 1986; Roshchupkin et al. 1979). The development of erythema therefore implies that enough ultraviolet damage has occurred that inflammatory pathways have been activated. Erythema is probably best thought of as a total failure of sun protection and is a marker for severe UV damage. It is now appreciated that there is a linkage between a history of repeated, severe sunburn and an increased risk for melanoma (Whiteman and Green 1994; MacKie and Aitchison 1982) and nonmelanoma skin cancer (Kricker et al. 1995; Gallagher et al. 1995; Kennedy et al. 2003; Naylor 1997).

Various strategies are followed for the protection of skin against UV-dependent damage. Limiting sun exposure, protective clothing, and the use of sunscreens are generally recommended. Systemic photoprotection through endogenous supply of nutritional bioactive agents such as carotenoids also provides important protection of the skin against the damaging effects of UV irradiation. However, far more research into the protective mechanisms of action of nutritional molecules on skin health is needed to fully appreciate and understand the long-term benefits of systemic photoprotection. In this chapter, we review the available evidence supporting a role for dietary carotenoids in systemic photoprotection.

5.5.2 Carotenoids are Scavengers of UV-Induced Reactive Oxygen Species that Damage DNA

Upon UV exposure, reactive oxygen species (ROS), which are constantly generated in skin, may be rapidly neutralized by nonenzymatic and enzymatic antioxidant substances. In consequence, cellular macromolecules (i.e., lipids, proteins, DNA) are protected from oxidation and a pro-oxidant/antioxidant balance is maintained, resulting in cell and tissue stabilization. If the antioxidant defense is exhausted, these ROS oxidize lipids, proteins, or DNA, leading to the formation of oxidized products such as lipid hydroperoxides, protein carbonyls, or 8-hydroxyguanosine, respectively (Beehler et al. 1992; Hu and Tappel 1992; Podda et al. 1998) and result in cell damage.

In recent years, carotenoids have gained considerable attention as a means to neutralize ROS (Mukhtar and Ahmad 1999). Because of their 11 conjugated double bonds, carotenoids such as β-carotene, lycopene, zeaxanthin, and lutein have potent antioxidant functionality and are among the most effective naturally occurring scavengers of single oxygen and peroxyl radicals (Cantrell et al. 2003; Di Mascio et al. 1989; Stahl and Sies 2003). Carotenoids are efficiently distributed to skin and, in consequence, might participate to the antioxidant capacity of the skin. Supplementation with lutein and zeaxanthin significantly decreases skin lipid peroxidation, as measured by malondialdehyde levels (Palombo et al. 2007; Morganti et al. 2002).

Considering that carotenoids exhibit high antioxidant activity in vitro, this property may be promising in neutralizing the ROS generated in vivo upon UV exposure. Although this effect is promising, it needs to be investigated in depth in human intervention studies with carotenoids and with validated markers of peroxidation (i.e., F-2 isoprostane for lipid peroxidation or the Comet assay for DNA peroxidation).

5.5.3 *Dietary Carotenoids Reduce Sunburn Development*

The sensitivity of an individual to erythematogenic UV exposure is determined by two methods: (1) the minimum erythema dose (MED), defined as the threshold dose required to cause perceptible reddening of the skin 24 h after exposure (Orentreich et al. 2001); or (2) the change of skin color assessed by chromametry (Stahl et al. 2006). With the latter technique, erythema is assessed by a change of chromametry a-values after and before irradiation (in daltons, or Da).

Decreasing Da values in comparison to those at week 0 (set to 100%) reflects protection against UV-induced erythema. In human intervention studies, the photoprotective effect of a nutritional intervention refers to UV-induced erythema as an early observable immediate response. It is measured either as increased MED or as a reduction of erythema intensity after UV exposure compared to baseline and to that of an unsupplemented group.

The efficacy of β-carotene in systemic photoprotection was investigated in seven human intervention studies (Garmyn et al. 1995; Mathews-Roth et al. 1972; McArdle et al. 2004; Lee et al. 2000; Stahl et al. 2000a; Heinrich et al. 2003; GOLLNICK et al. 1996). A meta-analysis (Kopcke and Krutmann 2008) of these seven studies demonstrated that β-carotene supplementation is efficient in providing protection against the development of sunburn (Table 5.1). The effect size is 0.8 SD [95% confidence interval (CI) 0.2–1.4]. The photoprotective effect is observed only in studies providing a daily dose of 20 mg for a minimum of 10 weeks of supplementation.

The efficacy of lycopene in systemic photoprotection was also investigated but to a lesser extent. Tomato paste provided a daily dietary intake of 16 mg lycopene. After 10 weeks of supplementation, dorsal erythema formation was 40% lower in the group that consumed tomato paste compared to the control group, although no difference between the groups was found after 4 weeks of treatment (Stahl et al. 2001). Results of another intervention study (Aust et al. 2005) supported this lycopene photoprotective effect. Daily intake of 10 mg of lycopene was provided from three sources: a synthetic lycopene, a tomato extract (Lyc-o-Mato), or a drink containing solubilized Lyc-o-Mato (Lyc-o-Guard-Drink). After 12 weeks of supplementation, erythema formation induced by UV exposure was significantly lower with each of the three lycopene treatments. Compared to week 0, the reductions were of 25%, 38%, and 48% for the synthetic lycopene, the tomato extract, and the drink, respectively. Stahl and coworkers (Stahl et al. 2006) also demonstrated this effect in a study where four lycopene sources—tomato paste, lycopene-rich carrot juice, lycopene drink, synthetic lycopene—were used to supply lycopene

Table 5.1 Carotenoids and skin erythema

Intervention dose (source)	Duration (weeks)	Results	Reference
β-Carotene			
90 mg/day	3	No protection	Garmyn et al. (1995)
15 mg/day (supplement)	8	No effect	McArdle et al. (2004)
180 mg/day	10	MED increased	Mathews-Roth et al. (1972)
24 mg/day (*D. salina*)	12	Erythema less pronounced	Stahl et al. (2000c)
24 mg/day (*D. salina*)	12	Erythema less pronounced	Heinrich et al. (2003)
30 mg/day (supplement)	12	Erythema less pronounced	Gollnick et al. (1996)
30 mg/day (*D. salina*)	8	MED increased	Lee et al. (2000)
60 mg/day (*D. salina*)	8		
90 mg/day (*D. salina*)	8		
Lycopene			
16 mg/day (tomato paste)	10–12	Erythema less pronounced	Stahl et al. (2001)
10 mg/day (synthetic lycopene)	12	Erythema less pronounced	Aust et al. (2005)
10 mg/day (tomato extract: Lyc-o-Mato)			
10 mg/day (beverage: Lyc-o-Guard-Drink)			
16 mg/day (tomato paste)	12	Erythema less pronounced except for synthetic lycopene	Stahl et al. (2006)
10.2 mg/day (synthetic lycopene)			
Carotenoid mix			
60 mg/day (β-carotene) + 90 mg/day (canthaxanthin)	4	No protection	Wolf et al. (1988)
8 mg/day (β-carotene) + 8 mg/day (lycopene) + 8 mg/day lutein	12	Erythema less pronounced	Heinrich et al. (2003)
5.1 mg/day (β-carotene, carrot juice) + 10 mg (lycopene, synthetic)	10	Erythema less pronounced except for synthetic lycopene	Stahl et al. (2006)

D. salina, *Dunaliella salina* (alga); *MED* minimum erythema dose

(8.2–16.0 mg/day) to volunteers over a period of 10–12 weeks. The paste, juice, and drink induced efficient photoprotection by 40%, 45%, and 50%, respectively. The synthetic lycopene showed a trend but it did not reach statistical significance (Stahl et al. 2006). Supplements derived from tomato-based products contain a number of other constituents, including other carotenoids such as phytofluene and phytoene, which are precursors of lycopene in the biosynthetic pathway. These compounds may well contribute to the photoprotective effects as they absorb in the UV range. This, then, may explain the absence of a photoprotective effect with the synthetic lycopene, as lycopene absorption into blood and its distribution to skin was good for the four products.

Human intervention studies with a mix of carotenoids (i.e., β-carotene, lycopene, lutein) confirmed the photoprotective effect of carotenoids (Heinrich et al. 2003). Erythema development was diminished in subjects whose diets were supplemented with β-carotene (24 mg/day) or a carotenoid mixture consisting of β-carotene, lutein, and lycopene (8 mg each/day) for 12 weeks (Stahl et al. 2001).

5.5.4 *Dietary Carotenoids and Skin Cancer*

Basal cell carcinoma (BCC) and squamous cell carcinoma (SCC) are the most commonly occurring skin cancers in white populations, and incidence rates have increased in Europe, the United States, and Australia (Christenson et al. 2005; de Vries et al. 2004; Staples et al. 2006). More than 1 million new cases of nonmelanoma skin cancers per year are being reported in the United States. According to estimates of the National Cancer Institute, 40–50% of Americans who live to age 65 develop skin cancer at least once, and the risk of developing additional tumors is high.

Since the 1980s, β-carotene has been proposed as a possible dietary preventive agent against cancer. In animal studies, β-carotene protects against skin cancer induced by chemicals and UV radiation (Krinsky 1989; Bollag 1970). At present, however, there is no clear evidence that carotenoids protect humans against skin cancer. Based on epidemiological studies, no association was found between dietary carotenoids and BCC (Fung et al. 2002; McNaughton et al. 2005) or SCC (Fung et al. 2003). In intervention studies, β-carotene supplementation failed to decrease the risk of nonmelanoma skin cancer among men with low baseline plasma β-carotene (Green et al. 1999; Greenberg et al. 1990; Darlington et al. 2003; Schaumberg et al. 2004). Moreover, β-carotene (30 mg/day) had no effect on the incidence of solar keratosis, a premalignant skin cancer, in a randomized controlled study in 1,600 participants (Darlington et al. 2003). These results were supported by another study showing that daily use of β-carotene (30 mg/day, $n=1621$) for 4.5 years did not reduce the incidence of BSC or SCC (Green et al. 1999).

5.6 Future Perspectives on Systemic Photoprotection by Carotenoids

Sunburn is just one phenomenon related to exposure to UV radiation. Erythema formation can be readily quantified and is being used successfully as a noninvasive parameter for assessing the biological response to UV exposure. Whether erythema formation is the most suitable surrogate endpoint for long-term degenerative diseases such as skin cancer (of various types), photodermatoses, or photoaging is not clear and needs to be scrutinized in further work. Therefore, protection against UV-induced erythema, or sunburn, does not necessarily mean protection against skin cancer.

Although evidence is available to support a role for dietary carotenoids (notably β-carotene, lutein, and lycopene) in protecting skin against sunburn development, it is not clear whether endogenous carotenoids can positively affect skin health by also interfering with UV-induced pathways that lead to DNA mutations, immune suppression, and skin cancer development. Moreover, there is much to be learned about their mechanisms of action. For example, it is unknown whether the bioefficacy of carotenoids in skin is due to a direct antioxidant function and/or independent of direct antioxidant chemistry, perhaps involving their ability to modulate gene expression by directly regulating cell signaling pathways. Thus, the modification of signaling cascades by nutrients is a developing area of research. Consequently, it may not be a prerequisite for a systemically photoprotective dietary bioactive to be present at or near a sensitive target site. Perhaps metabolites or oxidation products of parent carotenoids such as retinoids or apocarotenals are the ultimate active agents (Aust et al. 2003; Stahl et al. 2000b; Teicher et al. 1999). The mechanisms that lead to the incorporation of micronutrients such as carotenoids into different parts of the skin are also not yet known. The fact that there are large disparities between skin areas in terms of embedded micronutrients, as shown for instance in the high levels of β-carotene in the palm of the hand as compared to other skin areas (Stahl et al. 1998), suggest that different mechanisms are involved. It is noteworthy that although past studies used supplemental doses of carotenoids to achieve a photoprotective effect, it is not possible to achieve such high doses with an average number of servings of fruits and vegetables. Therefore, future studies should further explore the correlation between the average dietary carotenoid dose and its influence on systemic photoprotection.

The concept of endogenous dietary skin photoprotection is that it provides a systemic maintenance level at sensitive dermal and epidermal target layers—beyond those reached by topical and temporary coverage through the use of sunscreen. Although endogenous protection through individual dietary components (in terms of the sun protection factor) may be considerably lower than that achieved using topical sunscreens and take considerably longer to be reached, an increased lifelong overall systemic protection via dietary supply may contribute significantly to skin health and complement the use of sunscreens in protecting the skin against the damaging effects of solar UV exposure.

References

Alaluf S et al. Dietary carotenoids contribute to normal human skin color and UV photosensitivity. J Nutr. 2002;132(3):399–403.

Andersen PH et al. A time-correlation study of ultraviolet B-induced erythema measured by reflectance spectroscopy and laser Doppler flowmetry. Photodermatol Photoimmunol Photomed. 1991;8(3):123–8.

Aust O et al. Lycopene oxidation product enhances gap junctional communication. Food Chem Toxicol. 2003;41(10):1399–407.

Aust O et al. Supplementation with tomato-based products increases lycopene, phytofluene, and phytoene levels in human serum and protects against UV-light-induced erythema. Int J Vitam Nutr Res. 2005;75(1):54–60.

Bayerl C, Schwarz B, Jung EG. A three-year randomized trial in patients with dysplastic naevi treated with oral beta-carotene. Acta Derm Venereol. 2003;83(4):277–81.

Beehler BC et al. Formation of 8-hydroxydeoxyguanosine within DNA of mouse keratinocytes exposed in culture to UVB and H2O2. Carcinogenesis. 1992;13(11):2003–7.

Bollag W. Vitamin A and vitamin A acid in the prophylaxis and therapy of epithelial tumours. Int Z Vitaminforsch. 1970;40(3):299–314.

Borel P et al. Carotenoids in biological emulsions: solubility, surface-to-core distribution, and release from lipid droplets. J Lipid Res. 1996;37(2):250–61.

Britton G. Structure and properties of carotenoids in relation to function. FASEB J. 1995; 9(15):1551–8.

Bruch-Gerharz D et al. Accumulation of the xanthophyll lutein in skin amyloid deposits of systemic amyloidosis (al type). J Invest Dermatol. 2001;116(1):196–7.

Cantrell A et al. Singlet oxygen quenching by dietary carotenoids in a model membrane environment. Arch Biochem Biophys. 2003;412(1):47–54.

Castenmiller JJ et al. The food matrix of spinach is a limiting factor in determining the bioavailability of beta-carotene and to a lesser extent of lutein in humans. J Nutr. 1999;129(2):349–55.

Cavallo J, DeLeo VA. Sunburn. Dermatol Clin. 1986;4(2):181–7.

Chandler LA, Schwartz SJ. Isomerization and losses of trans-.beta.-carotene in sweet potatoes as affected by processing treatments. J Agric Food Chem. 1988;36(1):129–33.

Chitchumroonchokchai C, Failla ML. Hydrolysis of zeaxanthin esters by carboxyl ester lipase during digestion facilitates micellarization and uptake of the xanthophyll by Caco-2 human intestinal cells. J Nutr. 2006;136(3):588–94.

Christenson LJ et al. Incidence of basal cell and squamous cell carcinomas in a population younger than 40 years. JAMA. 2005;294(6):681–90.

Clydesdale GJ, Dandie GW, Muller HK. Ultraviolet light induced injury: immunological and inflammatory effects. Immunol Cell Biol. 2001;79(6):547–68.

Connor WE et al. The prime role of HDL to transport lutein into the retina: evidence from HDL-deficient WHAM chicks having a mutant ABCA1 transporter. Invest Ophthalmol Vis Sci. 2007;48(9):4226–31.

Darlington S et al. A randomized controlled trial to assess sunscreen application and beta carotene supplementation in the prevention of solar keratoses. Arch Dermatol. 2003;139(4):451–5.

Darvin ME et al. In vivo distribution of carotenoids in different anatomical locations of human skin: comparative assessment with two different Raman spectroscopy methods. Exp Dermatol. 2009;18(12):1060–3.

de Vries E et al. Rapid and continuous increases in incidence rates of basal cell carcinoma in the southeast Netherlands since 1973. J Invest Dermatol. 2004;123(4):634–8.

Deming DM, Baker DH, Erdman Jr JW. The relative vitamin A value of 9-cis beta-carotene is less and that of 13-cis beta-carotene may be greater than the accepted 50% that of all-trans beta carotene in gerbils. J Nutr. 2002;132(9):2709–12.

Di Mascio P, Kaiser S, Sies H. Lycopene as the most efficient biological carotenoid singlet oxygen quencher. Arch Biochem Biophys. 1989;274(2):532–8.

Dimitrov NV et al. Bioavailability of beta-carotene in humans. Am J Clin Nutr. 1988;48(2):298–304.

During A, Dawson HD, Harrison EH. Carotenoid transport is decreased and expression of the lipid transporters SR-BI, NPC1L1, and ABCA1 is downregulated in Caco-2 cells treated with ezetimibe. J Nutr. 2005;135(10):2305–12.

Edwards AJ et al. Alpha- and beta-carotene from a commercial puree are more bioavailable to humans than from boiled-mashed carrots, as determined using an extrinsic stable isotope reference method. J Nutr. 2002;132(2):159–67.

Ermakov IV et al. Noninvasive selective detection of lycopene and beta-carotene in human skin using Raman spectroscopy. J Biomed Opt. 2004;9(2):332–8.

Ermakov IV et al. Resonance Raman detection of carotenoid antioxidants in living human tissue. J Biomed Opt. 2005;10(6):064028.
Fisher GJ et al. Mechanisms of photoaging and chronological skin aging. Arch Dermatol. 2002;138(11):1462–70.
Fung TT et al. Vitamins and carotenoids intake and the risk of basal cell carcinoma of the skin in women (United States). Cancer Causes Control. 2002;13(3):221–30.
Fung TT et al. Vitamin and carotenoid intake and risk of squamous cell carcinoma of the skin. Int J Cancer. 2003;103(1):110–5.
Gallagher RP et al. Sunlight exposure, pigmentary factors, and risk of nonmelanocytic skin cancer I. Basal cell carcinoma. Arch Dermatol. 1995;131(2):157–63.
Garmyn M et al. Effect of beta-carotene supplementation on the human sunburn reaction. Exp Dermatol. 1995;4(2):104–11.
Gerster H. The potential role of lycopene for human health. J Am Coll Nutr. 1997;16(2):109–26.
Gollnick HP et al. Systemic beta carotene plus topical UV-sunscreen are an optimal protectionagainst harmful effects of natural UV-sunlight: results of the Berlin-Eilath study. Eur J Dermatol. 1996;6(3):200–5.
Green A et al. Daily sunscreen application and betacarotene supplementation in prevention of basal-cell and squamous-cell carcinomas of the skin: a randomised controlled trial. Lancet. 1999;354(9180):723–9.
Greenberg ER et al. A clinical trial of beta carotene to prevent basal-cell and squamous-cell cancers of the skin. The Skin Cancer Prevention Study Group. N Engl J Med. 1990;323(12):789–95.
Hammond BR, Wooten BR. Resonance Raman spectroscopic measurement of carotenoids in the skin and retina. J Biomed Opt. 2005;10(5):054002.
Hata TR et al. Non-invasive Raman spectroscopic detection of carotenoids in human skin. J Invest Dermatol. 2000;115(3):441–8.
Heinrich U et al. Supplementation with beta-carotene or a similar amount of mixed carotenoids protects humans from UV-induced erythema. J Nutr. 2003;133(1):98–101.
Helfrich YR, Sachs DL, Voorhees JJ. Overview of skin aging and photoaging. Dermatol Nurs. 2008;20(3):177–83. quiz 184.
Hesterberg K et al. Raman spectroscopic analysis of the increase of the carotenoid antioxidant concentration in human skin after a 1-week diet with ecological eggs. J Biomed Opt. 2009; 14(2):024039.
Hruza LL, Pentland AP. Mechanisms of UV-induced inflammation. J Invest Dermatol. 1993;100(1):35 S–41 S.
Hu ML, Tappel AL. Potentiation of oxidative damage to proteins by ultraviolet-A and protection by antioxidants. Photochem Photobiol. 1992;56(3):357–63.
Jalal F et al. Serum retinol concentrations in children are affected by food sources of beta-carotene, fat intake, and anthelmintic drug treatment. Am J Clin Nutr. 1998;68(3):623–9.
Kennedy C et al. The influence of painful sunburns and lifetime sun exposure on the risk of actinic keratoses, seborrheic warts, melanocytic nevi, atypical nevi, and skin cancer. J Invest Dermatol. 2003;120(6):1087–93.
Khachik F et al. Separation and quantitation of carotenoids in foods. Methods Enzymol. 1992;213:347–59.
Kopcke W, Krutmann J. Protection from sunburn with beta-Carotene–a meta-analysis. Photochem Photobiol. 2008;84(2):284–8.
Kostic D, White WS, Olson JA. Intestinal absorption, serum clearance, and interactions between lutein and beta-carotene when administered to human adults in separate or combined oral doses. Am J Clin Nutr. 1995;62(3):604–10.
Kricker A et al. Does intermittent sun exposure cause basal cell carcinoma? a case-control study in Western Australia. Int J Cancer. 1995;60(4):489–94.
Krinsky NI. Carotenoids as chemopreventive agents. Prev Med. 1989;18(5):592–602.
Lee J et al. Carotenoid supplementation reduces erythema in human skin after simulated solar radiation exposure. Proc Soc Exp Biol Med. 2000;223(2):170–4.

Livny O et al. Beta-carotene bioavailability from differently processed carrot meals in human ileostomy volunteers. Eur J Nutr. 2003;42(6):338–45.
MacKie RM, Aitchison T. Severe sunburn and subsequent risk of primary cutaneous malignant melanoma in Scotland. Br J Cancer. 1982;46(6):955–60.
Mathews-Roth MM et al. A clinical trial of the effects of oral beta-carotene on the responses of human skin to solar radiation. J Invest Dermatol. 1972;59(4):349–53.
Matsumura Y et al. Resistance of CD1d–/– mice to ultraviolet-induced skin cancer is associated with increased apoptosis. Am J Pathol. 2004;165(3):879–87.
McArdle F et al. Effects of oral vitamin E and beta-carotene supplementation on ultraviolet radiation-induced oxidative stress in human skin. Am J Clin Nutr. 2004;80(5):1270–5.
McNaughton SA, Marks GC, Green AC. Role of dietary factors in the development of basal cell cancer and squamous cell cancer of the skin. Cancer Epidemiol Biomarkers Prev. 2005;14(7):1596–607.
Meinke MC et al. Bioavailability of natural carotenoids in human skin compared to blood. Eur J Pharm Biopharm. 2010;76(2):269–74.
Melnikova VO, Ananthaswamy HN. Cellular and molecular events leading to the development of skin cancer. Mutat Res. 2005;571(1–2):91–106.
Metzger A et al. Antioxidant status and acute malaria in children in Kampala, Uganda. Am J Trop Med Hyg. 2001;65(2):115–9.
Micozzi MS et al. Carotenodermia in men with elevated carotenoid intake from foods and beta-carotene supplements. Am J Clin Nutr. 1988;48(4):1061–4.
Moodycliffe AM et al. Immune suppression and skin cancer development: regulation by NKT cells. Nat Immunol. 2000;1(6):521–5.
Morganti P et al. Role of topical and nutritional supplement to modify the oxidative stress. Int J Cosmet Sci. 2002;24(6):331–9.
Mukhtar H, Ahmad N. Cancer chemoprevention: future holds in multiple agents. Toxicol Appl Pharmacol. 1999;158(3):207–10.
National Academy of, S. and M. Institute of, Dietary Reference Intakes. Vitamin A, Vitamin K, Arsenic, Boron, Chromium, Copper, Iodine, Iron, Manganese, Molybdenum, Nickel, Silicon, Vanadium and Zinc. Vol. 4; 2001, National Academy Press. p. 1–61.
Naylor MF. Erythema, skin cancer risk, and sunscreens. Arch Dermatol. 1997;133(3):373–5.
Olson JA. Bioavailability of carotenoids. Arch Latinoam Nutr. 1999;49(3 Suppl 1):21 S–5 S.
Orentreich D et al. Sunscreens: practical applications. In: Giacomoni PU, editor. Sun Protection in Man. Amsterdam: Elsevier; 2001. p. 535–59.
Palombo P et al. Beneficial long-term effects of combined oral/topical antioxidant treatment with the carotenoids lutein and zeaxanthin on human skin: a double-blind, placebo-controlled study. Skin Pharmacol Physiol. 2007;20(4):199–210.
Podda M et al. UV-irradiation depletes antioxidants and causes oxidative damage in a model of human skin. Free Radic Biol Med. 1998;24(1):55–65.
Ramsay CA, Challoner AV. Vascular changes in human skin after ultraviolet irradiation. Br J Dermatol. 1976;94(5):487–93.
Reboul E et al. Lutein transport by Caco-2 TC-7 cells occurs partly by a facilitated process involving the scavenger receptor class B type I (SR-BI). Biochem J. 2005;387(Pt 2):455–61.
Ribaya-Mercado JD. Influence of dietary fat on beta-carotene absorption and bioconversion into vitamin A. Nutr Rev. 2002;60(4):104–10.
Rich GT et al. Solubilization of carotenoids from carrot juice and spinach in lipid phases: I. Modeling the gastric lumen. Lipids. 2003a;38(9):933–45.
Rich GT et al. Solubilization of carotenoids from carrot juice and spinach in lipid phases: II. Modeling the duodenal environment. Lipids. 2003b;38(9):947–56.
Richelle M et al. Lycopene isomerisation takes place within enterocytes during absorption in human subjects. Br J Nutr. 2010;103(12):1800–7.
Rodriguez-Amaya DB. Food carotenoids: analysis, composition and alterations during storage and processing of foods. Forum Nutr. 2003;56:35–7.
Roshchupkin DI, Pistsov MY, Potapenko AY. Inhibition of ultraviolet light-induced erythema by antioxidants. Arch Dermatol Res. 1979;266(1):91–4.

Ryan L, et al. Micellarisation of carotenoids from raw and cooked vegetables. Plant Foods Hum Nutr (Dordrecht, Netherlands). 2008;63(3):127–33.
Schaumberg DA et al. No effect of beta-carotene supplementation on risk of nonmelanoma skin cancer among men with low baseline plasma beta-carotene. Cancer Epidemiol Biomarkers Prev. 2004;13(6):1079–80.
Sies H, Stahl W. Carotenoids and UV protection. Photochem Photobiol Sci. 2004;3(8):749–52.
Stahl W, Sies H. Uptake of lycopene and its geometrical isomers is greater from heat-processed than from unprocessed tomato juice in humans. J Nutr. 1992;122(11):2161–6.
Stahl W, Sies H. Antioxidant activity of carotenoids. Mol Aspects Med. 2003;24(6):345–51.
Stahl W et al. cis-trans isomers of lycopene and beta-carotene in human serum and tissues. Arch Biochem Biophys. 1992;294(1):173–7.
Stahl W et al. Increased dermal carotenoid levels assessed by noninvasive reflection spectrophotometry correlate with serum levels in women ingesting Betatene. J Nutr. 1998;128(5):903–7.
Stahl W et al. Carotenoids and carotenoids plus vitamin E protect against ultraviolet light-induced erythema in humans. Am J Clin Nutr. 2000a;71(3):795–8.
Stahl W et al. Stimulation of gap junctional communication: comparison of acyclo-retinoic acid and lycopene. Arch Biochem Biophys. 2000b;373(1):271–4.
Stahl W et al. Carotenoids in human skin: noninvasive measurement and identification of dermal carotenoids and carotenol esters. Methods Enzymol. 2000c;319:494–502.
Stahl W et al. Dietary tomato paste protects against ultraviolet light-induced erythema in humans. J Nutr. 2001;131(5):1449–51.
Stahl W et al. Lycopene-rich products and dietary photoprotection. Photochem Photobiol Sci. 2006;5(2):238–42.
Staples MP et al. Non-melanoma skin cancer in Australia: the 2002 national survey and trends since 1985. Med J Aust. 2006;184(1):6–10.
Sundaresan PR, et al. Effects of dietary taurocholate, fat and protein on the storage and metabolism of dietary beta-carotene and alpha-tocopherol in ferrets. International Journal for Vitamin and Nutrition Research. Internationale Zeitschrift fur Vitamin-und Ernahrungsforschung. Journal international de vitaminologie et de nutrition 2005;75(2): 133–41.
Tang G et al. Gastric acidity influences the blood response to a beta-carotene dose in humans. Am J Clin Nutr. 1996;64(4):622–6.
Teicher VB et al. Biological activities of Apo-canthaxanthinoic acids related to gap junctional communication. Arch Biochem Biophys. 1999;365(1):150–5.
Thakkar SK et al. Impact of style of processing on retention and bioaccessibility of beta-carotene in cassava (Manihot esculanta, Crantz). J Agric Food Chem. 2009;57(4):1344–8.
Ullrich SE. Mechanisms underlying UV-induced immune suppression. Mutat Res. 2005;571(1–2): 185–205.
Unlu NZ et al. Carotenoid absorption from salad and salsa by humans is enhanced by the addition of avocado or avocado oil. J Nutr. 2005;135(3):431–6.
Vahlquist A et al. Vitamin A in human skin: II concentrations of carotene, retinol and dehydroretinol in various components of normal skin. J Invest Dermatol. 1982;79(2):94–7.
van den Berg H. Carotenoid interactions. Nutr Rev. 1999;57(1):1–10.
van den Berg H, van Vliet T. Effect of simultaneous, single oral doses of beta-carotene with lutein or lycopene on the beta-carotene and retinyl ester responses in the triacylglycerol-rich lipoprotein fraction of men. Am J Clin Nutr. 1998;68(1):82–9.
van het Hof KH et al. Bioavailability of lutein from vegetables is 5 times higher than that of beta-carotene. Am J Clin Nutr. 1999;70(2):261–8.
Whiteman D, Green A. Melanoma and sunburn. Cancer Causes Control. 1994;5(6):564–72.
Wingerath T, Sies H, Stahl W. Xanthophyll esters in human skin. Arch Biochem Biophys. 1998;355(2):271–4.
Wolf C, Steiner A, Honigsmann H. Do oral carotenoids protect human skin against ultraviolet erythema, psoralen phototoxicity, and ultraviolet-induced DNA damage? J Invest Dermatol. 1988;90(1):55–7.

Yonekura L, Nagao A. Intestinal absorption of dietary carotenoids. Mol Nutr Food Res. 2007;51(1):107–15.

Zanutto M.E, et al. Effect of citric pectin on beta-carotene bioavailability in rats. International Journal for Vitamin and Nutrition Research. Internationale Zeitschrift fur Vitamin-und Ernahrungsforschung. Journal international de vitaminologie et de nutrition 2002;72(4):199–203.

Zidichouski JA et al. Clinical validation of a noninvasive, Raman spectroscopic method to assess carotenoid nutritional status in humans. J Am Coll Nutr. 2009;28(6):687–93.

Chapter 6
Antioxidants and Skin

Juergen Lademann, Maxim E. Darvin, and Ulrike Heinrich

Core Messages

- Free radicals are of great importance to signaling processes in the human body. If their concentration exceeds a critical value, however, these highly reactive molecules can destroy cells or cell compartments. The reasons for enhanced radical formation in the human organism, specifically in the skin, are manifold. In addition to environmental factors, such as ultraviolet radiation of the sun and contact with environmental hazards, smoking and excessive alcohol consumption lead to the formation of free radicals in the skin. In addition, such formation can be stimulated by illness, insomnia, stress on the job or at home, and similar stress situations elsewhere.
- With the antioxidative network, the human body has developed a protective system against the harmful action of free radicals. The most important antioxidants in the human body, particularly in the skin, include vitamins A, C, E, and D; the carotenoids β-carotene, lycopene, and lutein; and polyphenols.
- Because most of these antioxidants cannot be produced by the human organism automatically, they must be taken in with food. Therefore, the antioxidants reflect both the lifestyle and physical condition of people. A diet rich in fruit and vegetables as well as stress reduction is the best prerequisite for a healthy organism and a reduction of the furrows and wrinkles associated with age.
- Modern methods for measuring antioxidants in human skin are discussed.

J. Lademann (✉)
Center of Experimental and Applied Cutaneous Physiology (CCP),
Department of Dermatology, Venerology and Allergology,
Charité – Universitätsmedizin Berlin, Germany
e-mail: juergen.lademann@charite.de

A. Pappas (ed.), *Nutrition and Skin: Lessons for Anti-Aging, Beauty and Healthy Skin*, DOI 10.1007/978-1-4419-7967-4_6,

6.1 Introduction

Free radicals are constantly formed in human skin as a result of cellular metabolism (Iannone et al. 1993), environmental conditions (Burke and Wei 2009), inflammation (Maeda and Akaike 1998), and sun irradiation (Zastrow et al. 2009). They play an important role in signaling processes inside the cell and between cellular systems (Droge 2002; Jackson 1999). The human body also utilizes free radicals to destroy viruses and bacteria (Rivero 2006; Akaike 2001). Although in the past comprehensive studies elucidated the radical production in human skin by ultraviolet (UV) light (Black 1987), recent investigations have shown that free radicals are formed in the skin in the visible and infrared ranges of the solar spectrum as well (Zastrow et al. 2009; Darvin et al. 2010a, b). A high concentration of free radicals can cause oxidative cell damage, skin aging, and cancer (Dhalla et al. 2000; Halliwell 2007; Palmer and Kitchin 2010; Sander et al. 2004). With the antioxidative network, the human body has developed a protective system against the destructive action of free radicals. In recent years it has been established that carotenoids function as antioxidants in lipophilic systems (Woodall et al. 1997). As several studies confirm, carotenoids not only protect algae and plants from intensive sunlight but function as photoprotective agents in humans (Aust et al. 2005; Heinrich et al. 2003).

Typical antioxidants in the human body are vitamins A, C, E, and D; the carotenoids β-carotene, lycopene, and lutein; and the polyphenols. As most of these antioxidants cannot be produced by the human organism automatically, so they must be taken in with food. In particular, fruit and vegetables contain high amounts of antioxidants (Darvin et al. 2007; Halvorsen et al. 2002). Their efficacy can be improved when they are combined with carotenoids and vitamins, such as α-tocopherol or vitamin C. This seems to be due to synergetic effects (Wrona et al. 2003). Furthermore, with regard to efficacy, the bioavailability of the agents plays a significant role. It is known that carotenoids are absorbed better from cooked food (e.g., tomato sauce/paste), and the presence of a lipid carrier (e.g., vegetable oil) results in greater bioavailability than that from raw vegetables (Gartner et al. 1997).

Because of the interaction of the antioxidants with the free radicals, these highly reactive molecules are neutralized before they can develop their damaging effect. Antioxidants form protective chains in the human skin and protect each other against the destructive action of free radicals. Recently, it was demonstrated that the carotenoids constitute markers of the complete antioxidative network (Darvin et al. 2008a).

If the concentration of free radicals in the body exceeds a critical threshold, cells or cell compartments are destroyed. This does not apply to molecules such as DNA alone but also to cellular structures such as the elastic fibers in elastin and collagen (Monboisse and Borel 1992; Kawaguchi et al. 1997). Thus, the free radicals comprise a decisive factor that expedites skin aging. The highly reactive molecules not only destroy the elastic fibers but prevent or reduce their regeneration. As a result, furrows and wrinkles occur in the skin. Thus, enhancement of the antioxidative network of human skin is recommended (Darvin et al. 2008b; Schroeder et al. 2008, 2010).

Although the interaction between antioxidants and free radicals has been a subject of intensive research for decades, this topic has recently received fresh impetus by the development of methods that permit antioxidants, specifically carotenoids, to be detected in the human skin noninvasively and online. The methods mainly applied here are reflection spectroscopy and resonance Raman spectroscopy (Darvin et al. 2005a; Stahl et al. 1998). In the past, such investigations had been possible only by performing biopsies or obtaining blood samples, which subsequently had to be subjected to time-consuming and expensive conventional methods, such as high-pressure liquid chromatography (HPLC) (Talwar et al. 1998).

In this chapter the results of various studies on the interaction of free radicals and antioxidants, specifically carotenoids, in the human skin are summarized and discussed.

6.2 Positive and Negative Effects of Antioxidants

Fruit and vegetables contain various antioxidants, which are also found in the human body and, more specifically, in the skin (Darvin et al. 2005b, 2007). Even in ancient times it had been known that intake of fruit and vegetables is important for staying healthy, but it had not been possible to demonstrate that healthy food positively influences the course of disease as well (Darvin et al. 2006).

Negative reports on the intake of fruits and vegetables do not exist from those days, except in rare cases of allergic reactions. Based on this positive experience, intensive efforts had been undertaken to extract antioxidants from fruit and vegetables or to produce them synthetically. Analyzing the results of studies on plant extracts and synthetic antioxidants, a wide range of applications have been found that are deemed to support medical therapies. These studies, however, provided very different results on the action of antioxidants. The therapy of cancer patients, for example, was supported by treatment with antioxidants. Some studies revealed that the patients of the verum group (therapy with products containing antioxidants) died earlier than those of the placebo group as a consequence of carotenoid uptake (Gallicchio et al. 2008). The results of these studies have severely damaged the reputation and importance of supportive treatment with antioxidants, specifically with β-carotene (Paolini et al. 2003; Zhang and Omaye 2001; Biesalski and Obermueller-Jevic 2001).

Contrary to medical reports, almost no adverse effects have been observed when antioxidants were used in the field of cosmetology (Cesarini et al. 2003; Dreher and Maibach 2001), except for isolated allergic reactions. This is due to the fact that the various antioxidants form protective chains in the human organism thus protecting themselves against the destructive action of free radicals. According to several studies, this protective network is more efficient as more components are contained in the antioxidant compound.

The best results were achieved when the composition and concentration of the antioxidants in food supplements were similar to those obtained by a healthy nutrition rich in fruit and vegetables. This is reasonable because the highest efficacy of the antioxidants could be obtained if they are administered at an amount similar to their physiological concentrations. For the application of single antioxidant components at high concentrations, there is a critical level. If this level is exceeded, these molecules can form radicals themselves. Unaware of this fact, various supportive treatments with antioxidants were carried out in the past based on the slogan, "the higher the level, the better the effect." Instead of reducing radical formation, this fatal error stimulated such formation, especially when single components were administered at high concentrations, which affected the course of the disease negatively (Darvin et al. 2006).

The situation is quite different when it comes to cosmetic treatments. In most cases, antioxidants are used as antiaging strategies; that is, the products are designed for long-term application. Most cosmetic products, therefore, contain significantly lower concentrations of antioxidants than products for supportive treatment with antioxidants. Based on the analysis of various studies it can be stated that the action of the antioxidants is more efficient as their composition and concentration correspond to the physiological conditions in the human body. In other words: The closer the supplementation resembles the intake of natural foods, the more efficient is the protective action of the antioxidative network. Therefore, it should be emphasized that with the exception of isolated allergic reactions, no negative actions of antioxidants have been reported from the intake of fruit and vegetables. Excessive uptake of antioxidants is prevented by a natural saturation effect of the human organism. However, systemically applied antioxidants or antioxidant compounds in the form of tablets or drinks may contain much higher amounts of antioxidants than the ones that are usually taken up daily from food.

For useful supplementation with antioxidants one needs to pay attention to good compatibility, high bioavailability, and synergies (i.e., with vitamins). Moreover, the intake must be adjusted to daily needs.

6.3 Polyphenols: Cocoa and Green Tea

Several antioxidants given systemically or topically as enriched food and supplements have been demonstrated to provide photoprotection. They include flavanols, carotenoids, tocopherols, and vitamin C (Dinkova-Kostova 2008; Sies and Stahl 2004; Nichols and Katiyar SK: Skin photoprotection by natural polyphenols: anti-inflammatory, antioxidant, DNA repair mechanisms. Arch Dermatol Res. 07 November 2009).

Flavonoids comprise a group of secondary plant constituents widespread in nature and available from dietary sources such as cocoa, green tea, soy, berries, or other fruit (Manach et al. 2004). Flavonoid-containing phytomedicines are used as antiinflammatory and antiallergic remedies, and a flavonoid-rich diet is suggested to play a role in the prevention of several kinds of cancer and cardiovascular disorders. Many of the alleged effects have been linked to the antioxidant properties of flavonoids, but they also exhibit other biological activities (Stevenson and Hurst 2007).

A study with high- and low-flavanol cocoa products demonstrated that ingestion of dietary flavanols from cocoa contributes to endogenous photoprotection and improves dermal blood circulation (Heinrich et al. 2006a; Neukam et al. 2007). Also, cocoa flavanols affect the cosmetically relevant parameters of skin surface and hydration. Flavonoids are used in common cosmetics primarily for their antioxidant and soothing actions.

In vitro and animal studies have provided evidence that tea flavanols, when applied orally or topically, ameliorate adverse skin reactions following UV exposure, including skin damage, erythema, and lipid peroxidation (Katiyar et al. 2000). Topical application of green tea polyphenols to human skin have inhibited the UVB-induced erythema response and decreased formation of cyclobutane pyrimidine dimers in skin, found in both epidermis and dermis (Hsu 2005). Pretreatment of skin with green tea extracts led to a lower number of sunburned cells after exposure to solar-simulated radiation with a minimum erythema dose (MED) of 2 and protected epidermal Langerhans cells from UV damage. Green tea polyphenols applied topically have been shown to protect DNA and prevent other damaging effects of UV light such as sunburn response, immunosuppression, and photoaging of the skin (Yusuf et al. 2007). Given the results of previous studies, there is strong evidence to support the concept that the consumption of dietary flavonoids from tea may confer photoprotection and improve skin quality.

In a previous study (Heinrich et al. 2006a) two groups of volunteers consumed either high-polyphenol (326 mg/day) or low-polyphenol (27 mg/day) cacao powder over a period of 12 weeks. Photoprotective effects and parameters of skin condition were measured. UV-induced erythema following exposure of selected skin areas to 1.25 MED was significantly decreased in the high-polyphenol group after 6 and 12 weeks of treatment; no change was found in the low-polyphenol group. Ingestion of high-polyphenol cacao powder led to an increase in the blood flow of cutaneous and subcutaneous tissues, whereas no change in blood flow was found in the low-polyphenol group. Supplementation with high-polyphenol cacao powder results in a significant increase in skin density and thickness. Neither parameter was affected in the low-polyphenol group. Evaluation of the skin surface showed a significant decrease of skin roughness in the high-polyphenol group, whereas no change was found in the low-polyphenol group comparing the starting values with weeks 6 and 12. No difference was found in the low-polyphenol group. The presented data show that ingestion of dietary polyphenols from cacao contributes to endogenous photoprotection and improves blood circulation in skin. Cacao polyphenols further affect cosmetically relevant parameters of skin surface and hydration.

In a second study (Neukam et al. 2007), the acute effects of a single dose of cocoa rich in flavanols on dermal microcirculation were investigated. In a crossover design study, 10 healthy women ingested a cocoa drink with a high (329 mg) or low (27 mg) content of flavanols. The major flavanol monomer in both drinks was (−)-epicatechin: 61 mg/dl in the high-flavanol product and 6.6 mg/dl in the low flavanol product. Dermal blood flow and oxygen saturation were examined by laser Doppler flowmetry and spectroscopically at 1 mm skin depth at 0, 1, 2, 4, and 6 h. At the same time points, the plasma levels of total epicatechin (free compound plus conjugates) were measured by means of HPLC. Subsequent to the intake of high-flavanol cocoa, dermal blood flow was significantly increased (by 1.7-fold) at 2 h and oxygen saturation was elevated 1.8-fold. No statistically significant changes

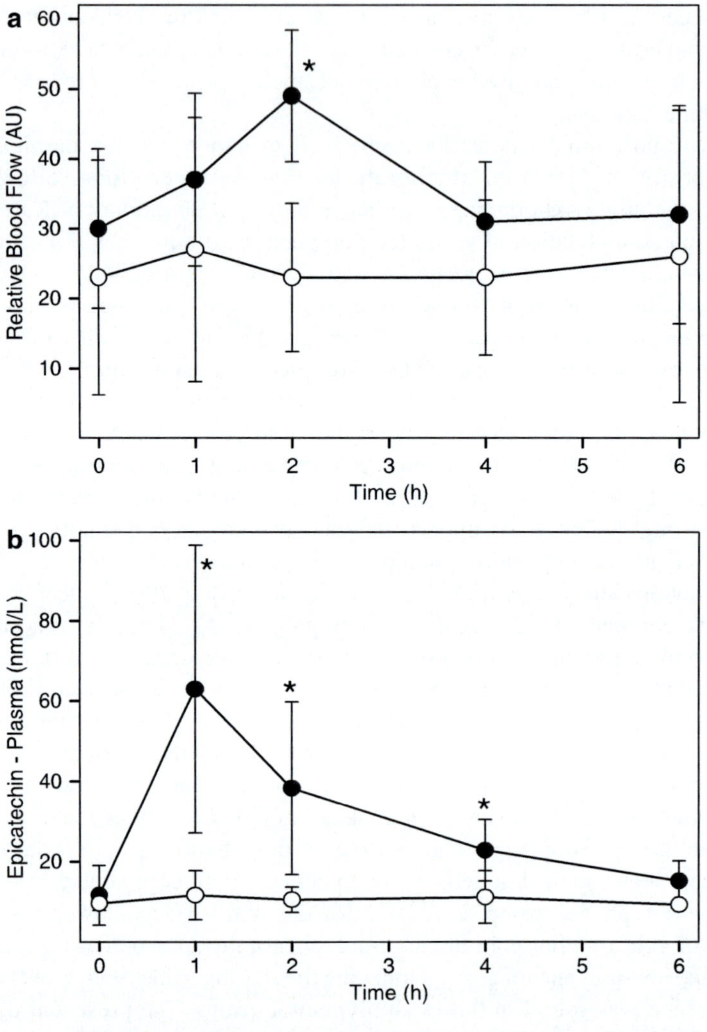

Fig. 6.1 Effect of high an low flavanol cocoa on cutaneous blood flow and plasma levels of total epicatechin. (**a**) Peripheral blood flow in skin (1 mm depth) after ingestion of a single dose of high-flavanol (*filled circles*) or low-flavanol (*open circles*) cocoa drink ($n=10$). (**b**) Plasma levels of total epicatechin (free epicatechin plus glucuronate and sulfate conjugates) after ingestion of a single dose of high-flavanol (*filled circles*) or low-flavanol (*open circles*) cocoa drink ($n=10$). *Significantly different from the low-flavanol group

were found with intake of low-flavanol cocoa. Maximum plasma levels of total epicatechin were observed 1 h after ingestion of the high-flavanol cocoa drink, 11.6 ± 7.4 nmol/l at baseline and 62.9 ± 35.8 nmol/l at 1 h. No change of total epicatechin was found in the low-flavanol group (Fig. 6.1a, b). The results led to the conclusion that flavanol-rich cocoa acutely improves dermal blood flow and oxygen saturation. Dietary flavanols may contribute to the maintenance of skin health and may influence skin appearance.

In a recent study, 60 volunteers were randomized into an intervention or control group. Over a period of 12 weeks, subjects consumed a drink with green tea polyphenols. Photoprotection parameters, skin structure, and function were measured at baseline (weeks 0, 6, and 12. Following exposure of the skin areas to 1.25 MED of radiation from a solar simulator, UV-induced erythema was significantly decreased in the intervention group. Skin structural parameters that were positively influenced included elasticity, density, roughness and scaling, and water homeostasis. Long-term intake of the green tea polyphenol beverage increased blood flow in the skin (article in preparation).

6.4 Antioxidants Protect Human Skin from Premature Aging

Various studies in which hundreds of volunteers had been investigated with the noninvasive Raman spectrometer disclosed that those volunteers who exhibited a high carotenoid level looked younger than they actually were. This subjective evaluation was subsequently substantiated by an objective study. For this purpose, healthy volunteers aged 40–50 years were selected who had not changed their lifestyle in recent years. Using a noninvasive optical method, the skin surface structure of these volunteers was assessed for skin aging. The measurements were made on the forehead as this is an area exposed to light. The carotenoid concentration was measured on the same area. An optical skin profile analyzer Primos (GFM Messtechnik, Teltow, Germany) was applied for these measurements (Roques et al. 2003). The measured area was 1×2 cm. Based on the skin surface measurements, skin roughness was measured (Lademann et al. 2008). The results showed a clear correlation between the antioxidant level in the human skin and skin roughness, which is determined by the depth and density of the furrows and wrinkles. The correlation factor R^2 was 0.7 (Darvin et al. 2008b). This result is not surprising because solar UV radiation is the main reason for destruction of the collagen and elastin fibers. Premature skin aging is the consequence. A high concentration of antioxidants, including carotenoids, in human skin can efficiently neutralize free radicals before they can develop their harmful effect. The achieved results correlate with those of Heinrich et al., who demonstrated that skin roughness is reduced after systemic application of antioxidants (Heinrich et al. 2006b).

Another significant approach to the antiaging concept is the improvement of elasticity. However, this particular aspect can be influenced to only a limited extent by external care. Nevertheless, studies with antioxidants show positive developments. The increase in skin density and thickness is caused by a fortifying effect on the collagen structure of the skin, which in turn has a positive effect on its elasticity and vigor. The ingestion of a nutritional supplement with various natural carotenoids, combined with selenium and vitamin E, over a period of 12 weeks resulted in a significant increase in skin density and thickness, which became apparent during an ultrasonography examination. At the same time, the surface structure of the skin was positively influenced (Heinrich et al. 2006b). Figures 6.2 and 6.3 show these observations.

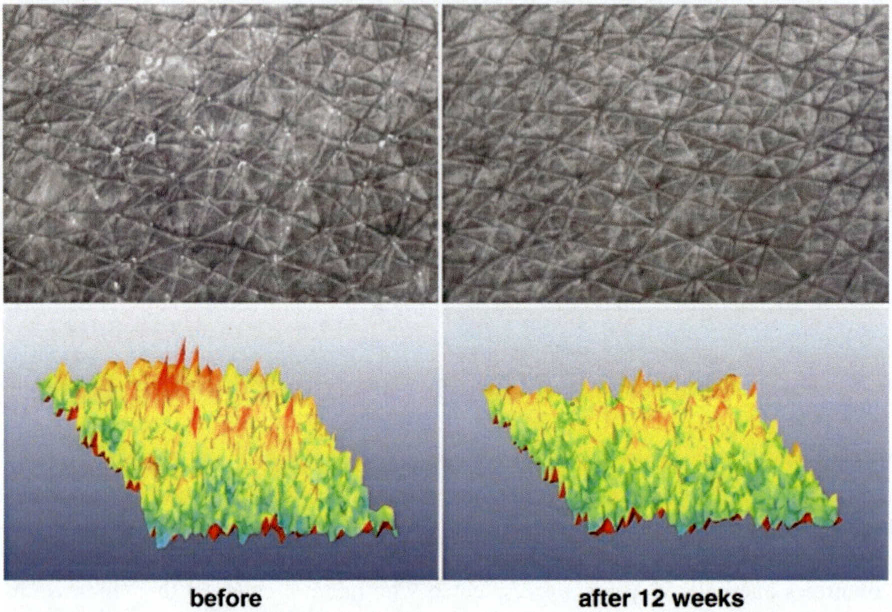

Fig. 6.2 Skin surface before and after 12 weeks of supplementation with a dietary supplement with mixed carotenoids (Heinrich et al. 2006b)

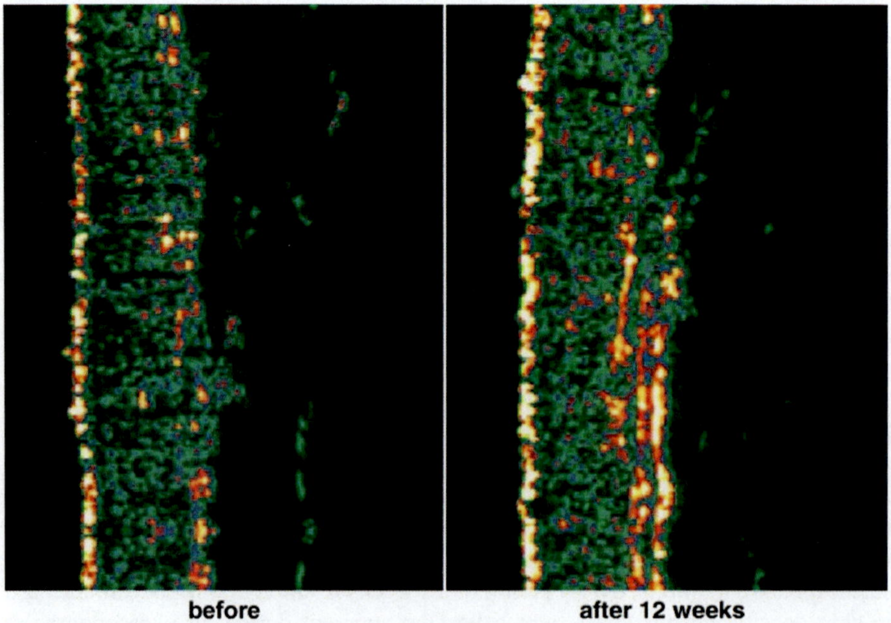

Fig. 6.3 Skin density and thickness measurement with ultrasonography (B-Scan) before and after 12 weeks of supplementation with a dietary supplement containing mixed carotenoids (Heinrich et al. 2006b)

Thus, a healthy lifestyle, including a generous uptake of fruit and vegetables, represents the best preventive strategy against premature skin aging.

Take Home Messages

- It has been shown that the carotenoid level reflects the lifestyle of each individual volunteer.
- High antioxidant—specifically carotenoid—levels in human skin can only be obtained by ingesting healthy food and the absence of stress. Stressful situations such as disease, exposure to intensive solar radiation, consumption of alcohol, and smoking reduce the carotenoid concentration in human skin. A high carotenoid level is the best prevention strategy against skin aging.
- Our daily diet does not always guarantee a sufficient supply of antioxidants and vitamins despite a large supply of fresh fruits and vegetables. Thus, a combination of external skin care together with internal supplementation, in addition to a healthy diet, is an excellent basis on which to guarantee the optimal supply of nutrients for the skin. Fast results should not be expected.
- All of the studies have shown that long-term adjustments are necessary, just as is the case with nutrition in general. Only regular ingestion of essential active agents in combination with adequate care for the respective skin type will lead to the desired success. Therefore, ingestion of dietary supplements, at least temporarily, can be conducive to increasing the photoprotection of the body and improving skin characteristics. Avoiding undesirable side effects as well as toxicological risks must have the highest priority when it comes to the development of supplements, especially if long-term ingestion is intended. Dietary supplements should never be seen as compensation for an unhealthy diet.

References

Akaike T. Role of free radicals in viral pathogenesis and mutation. Rev Med Virol. 2001;11:87–101.

Aust O, Stahl W, Sies H, Tronnier H, Heinrich U. Supplementation with tomato-based products increases lycopene, phytofluene, and phytoene levels in human serum and protects against UV-light-induced erythema. Int J Vitam Nutr Res. 2005;75:54–60.

Biesalski HK, Obermueller-Jevic UC. UV light, beta-carotene and human skin-beneficial and potentially harmful effects. Arch Biochem Biophys. 2001;389:1–6.

Black HS. Potential involvement of free radical reactions in ultraviolet light-mediated cutaneous damage. Photochem Photobiol. 1987;46:213–21.

Burke KE, Wei H. Synergistic damage by UVA radiation and pollutants. Toxicol Ind Health. 2009;25:219–24.

Cesarini JP, Michel L, Maurette JM, Adhoute H, Bejot M. Immediate effects of UV radiation on the skin: modification by an antioxidant complex containing carotenoids. Photodermatol Photoimmunol Photomed. 2003;19:182–9.

Darvin ME, Gersonde I, Albrecht H, Gonchukov SA, Sterry W, Lademann J. Determination of beta carotene and lycopene concentrations in human skin using resonance Raman spectroscopy. Laser Phys. 2005a;15:295–9.
Darvin ME, Gersonde I, Meinke M, Sterry W, Lademann J. Non-invasive in vivo determination of the carotenoids beta-carotene and lycopene concentrations in the human skin using the Raman spectroscopic method. J Phys D Appl Phys. 2005b;38:2696–700.
Darvin M, Zastrow L, Sterry W, Lademann J. Effect of supplemented and topically applied antioxidant substances on human tissue. Skin Pharmacol Physiol. 2006;19:238–47.
Darvin ME, Gersonde I, Albrecht H, Sterry W, Lademann J. Resonance Raman spectroscopy for the detection of carotenolds in foodstuffs. Influence of the nutrition on the antioxidative potential of the skin. Laser Phys Lett. 2007;4:452–6.
Darvin ME, Patzelt A, Knorr F, Blume-Peytavi U, Sterry W, Lademann J. One-year study on the variation of carotenoid antioxidant substances in living human skin: influence of dietary supplementation and stress factors. J Biomed Opt. 2008a;13:044028.
Darvin M, Patzelt A, Gehse S, Schanzer S, Benderoth C, Sterry W, et al. Cutaneous concentration of lycopene correlates significantly with the roughness of the skin. Eur J Pharm Biopharm. 2008b;69:943–7.
Darvin ME, Haag S, Meinke M, Zastrow L, Sterry W, Lademann J. Radical production by infrared A irradiation in human tissue. Skin Pharmacol Physiol. 2010a;23:40–6.
Darvin ME, Haag SF, Lademann J, Zastrow L, Sterry W, Meinke MC. Formation of free radicals in human skin during irradiation with infrared light. J Invest Dermatol. 2010b;130:629–31.
Dhalla NS, Temsah RM, Netticadan T. Role of oxidative stress in cardiovascular diseases. J Hypertens. 2000;18:655–73.
Dinkova-Kostova AT. Phytochemicals as protectors against ultraviolet radiation: versatility of effects and mechanisms. Planta Med. 2008;74(13):1548–59.
Dreher F, Maibach H. Protective effects of topical antioxidants in humans. Curr Probl Dermatol. 2001;29:157–64.
Droge W. Free radicals in the physiological control of cell function. Physiol Rev. 2002;82:47–95.
Gallicchio L, Boyd K, Matanoski G, Tao XG, Chen L, Lam TK, et al. Carotenoids and the risk of developing lung cancer: a systematic review. Am J Clin Nutr. 2008;88:372–83.
Gartner C, Stahl W, Sies H. Lycopene is more bioavailable from tomato paste than from fresh tomatoes. Am J Clin Nutr. 1997;66:116–22.
Halliwell B. Biochemistry of oxidative stress. Biochem Soc Trans. 2007;35:1147–50.
Halvorsen BL, Holte K, Myhrstad MC, Barikmo I, Hvattum E, Remberg SF, et al. A systematic screening of total antioxidants in dietary plants. J Nutr. 2002;132:461–71.
Heinrich U, Gartner C, Wiebusch M, Eichler O, Sies H, Tronnier H, et al. Supplementation with beta-carotene or a similar amount of mixed carotenoids protects humans from UV-induced erythema. J Nutr. 2003;133:98–101.
Heinrich U, Neukam K, Tronnier H, Sies H, Stahl W. Long-term ingestion of high flavanol cocoa provides photoprotection against UV-induced erythema and improves skin condition in women. J Nutr. 2006a;136(6):1565–9.
Heinrich U, Tronnier H, Stahl W, Bejot M, Maurette JM. Antioxidant supplements improve parameters related to skin structure in humans. Skin Pharmacol Physiol. 2006b;19:224–31.
Hsu S. Green tea and the skin. J Am Acad Dermatol. 2005;52(6):1049–59.
Iannone A, Marconi A, Zambruno G, Giannetti A, Vannini V, Tomasi A. Free-radical production during metabolism of organic hydroperoxides by normal human keratinocytes. J Invest Dermatol. 1993;101:59–63.
Jackson MJ. Free radicals in skin and muscle: damaging agents or signals for adaptation? Proc Nutr Soc. 1999;58:673–6.
Katiyar SK, Perez A, Mukhtar H. Green tea polyphenol treatment to human skin prevents formation of ultraviolet light B-induced pyrimidine dimers in DNA. Clin Cancer Res. 2000; 6(10):3864–9.

Kawaguchi Y, Tanaka H, Okada T, Konishi H, Takahashi M, Ito M, et al. Effect of reactive oxygen species on the elastin mRNA expression in cultured human dermal fibroblasts. Free Radic Biol Med. 1997;23:162–5.
Lademann J, Patzelt A, Darvin M, Richter H, Antoniou C, Sterry W, et al. Application of optical non-invasive methods in skin physiology. Laser Phys Lett. 2008;5:335–46.
Maeda H, Akaike T. Nitric oxide and oxygen radicals in infection, inflammation, and cancer. Biochemistry (Mosc). 1998;63:854–65.
Manach C, Scalbert A, Morand C, Remesy C, Jimenez L. Polyphenols: food sources and bioavailability. Am J Clin Nutr. 2004;79(5):727–47.
Monboisse JC, Borel JP. Oxidative damage to collagen. EXS. 1992;62:323–7.
Neukam K, Stahl W, Tronnier H, Sies H, Heinrich U. Consumption of flavanol-rich cocoa acutely increases microcirculation in human skin. Eur J Nutr. 2007;46(1):53–6.
Nichols JA, Katiyar SK. Skin photoprotection by natural polyphenols: anti-inflammatory, antioxidant and DNA repair mechanisms. Arch Dermatol Res. 07 November 2009. doi: 10.1007/s00403-00901001-1. PubMed PMID: 1989885
Palmer DM, Kitchin JS. Oxidative damage, skin aging, antioxidants and a novel antioxidant rating system. J Drugs Dermatol. 2010;9:11–5.
Paolini M, bdel-Rahman SZ, Sapone A, Pedulli GF, Perocco P, Cantelli-Forti G, et al. beta-Carotene: a cancer chemopreventive agent or a co-carcinogen? Mutat Res. 2003;543:195–200.
Rivero A. Nitric oxide: an antiparasitic molecule of invertebrates. Trends Parasitol. 2006;22:219–25.
Roques C, Teot L, Frasson N, Meaume S. PRIMOS: an optical system that produces three-dimensional measurements of skin surfaces. J Wound Care. 2003;12:362–4.
Sander CS, Chang H, Hamm F, Elsner P, Thiele JJ. Role of oxidative stress and the antioxidant network in cutaneous carcinogenesis. Int J Dermatol. 2004;43:326–35.
Schroeder P, Lademann J, Darvin ME, Stege H, Marks C, Bruhnke S, et al. Infrared radiation-induced matrix metalloproteinase in human skin: implications for protection. J Invest Dermatol. 2008;128:2491–7.
Schroeder P, Calles C, Benesova T, Macaluso F, Krutmann J. Photoprotection beyond ultraviolet radiation–effective sun protection has to include protection against infrared A radiation-induced skin damage. Skin Pharmacol Physiol. 2010;23:15–7.
Sies H, Stahl W. Nutritional protection against skin damage from sunlight. Annu Rev Nutr. 2004;24:173–200.
Stahl W, Heinrich U, Jungmann H, von LJ, Schietzel M, Sies H, et al. Increased dermal carotenoid levels assessed by noninvasive reflection spectrophotometry correlate with serum levels in women ingesting Betatene. J Nutr. 1998;128:903–7.
Stevenson DE, Hurst RD. Polyphenolic phytochemicals–just antioxidants or much more? Cell Mol Life Sci. 2007;64(22):2900–16.
Talwar D, Ha TK, Cooney J, Brownlee C, O'Reilly DS. A routine method for the simultaneous measurement of retinol, alpha-tocopherol and five carotenoids in human plasma by reverse phase HPLC. Clin Chim Acta. 1998;270:85–100.
Woodall AA, Lee SWM, Weesie RJ, Jackson MJ, Britton G. Oxidation of carotenoids by free radicals: relationship between structure and reactivity. Biochim Biophys Acta. 1997;1336:33–42.
Wrona M, Korytowski W, Rozanowska M, Sarna T, Truscott TG. Cooperation of antioxidants in protection against photosensitized oxidation. Free Radic Biol Med. 2003;35:1319–29.
Yusuf N, Irby C, Katiyar SK, Elmets CA. Photoprotective effects of green tea polyphenols. Photodermatol Photoimmunol Photomed. 2007;23(1):48–56.
Zastrow L, Groth N, Klein F, Kockott D, Lademann J, Renneberg R, et al. The missing link–light-induced (280–1,600 nm) free radical formation in human skin. Skin Pharmacol Physiol. 2009;22:31–44.
Zhang P, Omaye ST. Antioxidant and prooxidant roles for beta-carotene, alpha-tocopherol and ascorbic acid in human lung cells. Toxicol In Vitro. 2001;15:13–24.

Chapter 7
Minerals and the Skin

Petra Winkler

Core Messages

- Several minerals such as zinc, iron, and copper have an impact on the skin.
- Adequate intake of minerals positively influences the overall appearance of the skin.
- Deficiency of certain minerals results in certain manifestations in skin, hair, and nails and it is associated with skin diseases.
- Dermatological effects are observed when the minerals are either orally supplemented or topically applied.
- Mineral supplements should be taken with care because excessive intake of minerals can lead to intoxication.

7.1 Introduction

Although only approximately 4% of the human body mass consists of minerals, they play a significant role. Minerals are involved in maintenance of electrical neutrality, osmotic pressure, solubility, buffer systems, nervous conditions, and metabolism via constituents of enzymes; they are also components of bones and teeth (Anke et al. 1999). Some minerals are also important to the appearance and health of the skin (Table 7.1). Minerals are classified as trace elements (each <0.1% of body weight) and major elements (each >0.1% of body weight). Because these substances take an active part in the metabolism, humans need a regular supply of these essential nutrients.

P. Winkler (✉)
J&J, Medical Affairs, EMEA, Neuss, Germany
e-mail: pwinkle2@its.jnj.com

A. Pappas (ed.), *Nutrition and Skin: Lessons for Anti-Aging, Beauty and Healthy Skin*,
DOI 10.1007/978-1-4419-7967-4_7,

Table 7.1 Minerals for treatment of skin conditions or prevention and therapy of skin diseases

Condition/ disease	Mineral	Recommended dose	Function
Acne	Zinc	200–600 mg topically applied	Antiinflammatory, antibacterial, reduction of sebum secretion
	Chromium	400 mg	Reduction of inflammation and severity of disease pattern
	Selenium (in combination with tocopherol)	400 μg (20 mg)	
Brittle hair and nails	Silicon	10 mg	
	Iron		Seen with iron deficiency
Diaper dermatitis	Zinc	Ointments – topically applied	Antiinflammatory, antimicrobial
Dandruff	Zinc-pyrithione Selenium	Shampoos, etc. – topically applied	Antifungal, anti-inflammatory, reduction of sebum secretion
Hair shedding	Iron (in combination with L-lysine)	72 mg (1.5 g)	
Herpes simplex infection	Zinc	15–30 mg for prevention 60–100 mg for active infection	Antiviral effect, may alleviate severity and frequency of outbreaks
Photodamage, antiaging	Copper	Copper-containing creams	Collagen and elastin synthesis
	Silicon	10 mg	
Pruritus	Iron		Seen with iron deficiency
Psoriasis	Combination of selenium and zinc	Zn 200 μg, Se 50 mg	Alleviation of inflammation, itching, and redness
	Immersion in and drinking of selenium-rich spa	70 μg/L	Improvement in psoriatic plaques
Wound healing	Zinc		Anti-inflammatory, cell communication, protein synthesis

7.2 Zinc

With approximately 1.5–2.5 g, zinc is the second most abundant trace element in the human body. It can be found in all living cells and body fluids. Amounts of 6–20% of the body stores have been described to be found in the skin, mainly as metalloenzymes, where zinc is needed because of the highly proliferative nature of the tissue. A continuous zinc supply is necessary as the body is not able to store the mineral and make use of it in case of an impending deficiency. Hence, a reduction of the alimentary supply leads rapidly to zinc deficiency (King et al. 2000; Henzel et al. 1970; Schwartz et al. 2005).

Table 7.2 Recommended daily intake for zinc for adults

Healthy people/condition	RDA (mg)
Men	10–11
Women	7–8
Pregnancy	10–11
Lactation	11–12

Data are from references Food and Nutrition Board, Institute of Medicine, National Academy of Sciences (2001) and D-A-CH (2008)
RDA, recommended daily allowance

7.2.1 Functions

- Cofactor in more than 200 enzymes and metalloproteins: e.g., RNA polymerase, alcohol dehydrogenase, various enzymes of DNA synthesis
- Involved in the metabolism of numerous hormones (e.g., growth hormone, thyroxine, insulin, sexual hormones) and the metabolism of proteins, fat, and carbohydrates
- Cell growth and differentiation
- Stabilization of the structures of DNA, RNA, and membranes
- Involvement in collagen synthesis and thereby essential for the construction and degradation of connective tissue, ligaments, and tendons
- Production and regulation of cellular and humoral immune reactions
- Cytoprotection against organic toxins, heavy metals, radiation, and endotoxins from pathogen bacteria
- Antioxidant (as part of the copper/zinc/superoxide dismutase combination)

Good sources of zinc are beef, pork, poultry, eggs, milk, cheese, whole wheat bread, bran, wheat germ, and oatmeal. However, the actual amount of zinc in zinc-rich food (e.g., whole wheat) may decrease due to processing (e.g., flour depending on the degree of fineness).

The utilization of zinc from food is of high importance to meet the daily zinc demand. Approximately 10–30% of the zinc consumed in the diet can be absorbed by the human body (Table 7.2). In general, the absorption from food derived from animals is better than that from vegetables. Chelating agents, such as histidine and cysteine, may increase absorption, whereas phytate and phosphate build complexes with zinc and therefore decrease absorption. A high content of calcium, stressful situations, surgical interventions, and parasitic diseases may also have a negative influence on the absorption rate (Food and Nutrition Board, Institute of Medicine, National Academy of Sciences 2001; D-A-CH 2008).

The toxicological threshold of zinc is high. Acute zinc intoxication may occur after consumption of acidic food or water from galvanized containers. Acute intoxication with, for example, 2 g of zinc causes gastrointestinal disorders, circulatory disorders, and fever. Chronic intoxication with >110 mg/day causes hypochromic anemia and neutropenia probably due to interaction with copper. Also, a short-term supply of about 50 mg of zinc per day leads to interference with iron and copper metabolism (Wolffram 1996). Therefore, zinc intake of more than 30 mg/day is not recommended, except for therapeutic reasons (Ryan and Goldsmith 1996). For the latter, gastrointestinal side effects can be somewhat reduced by consuming zinc directly after meals. Because zinc decreases the absorption of copper, 1–2 mg of copper supplementation may be recommended in patients on chronic zinc therapy to prevent copper deficiency (Bowe and Shalita 2008). Cutaneous manifestations of zinc intoxication have not been reported (Ryan and Goldsmith 1996).

7.2.2 *Deficiency*

Although plasma zinc concentration is the most reliable measure of the total bodily storage, it is not always an accurate reflection. In fact, there is no test that can make this determination with accuracy. The range of normal values is broad (90 ± 20 μg/dL). Zinc levels in hair also vary greatly and can be elevated falsely by traces of tightly bound zinc if zinc shampoo has been used (Neldner 1984).

In addition to the autosomal recessive hereditary disorder acrodermatitis enteropathica, which is characterized by the inability to absorb zinc from the intestine, zinc deficiency may occur in malabsorption syndromes (e.g., Crohn's disease or celiac disease), with parenteral nutrition, during treatment with chelating agents, or in cases of severe skin burns (Neldner 1993).

Zinc deficiency due to malnutrition was discovered in rural areas of the Middle East, where diets consist mainly of cereals, which contain phytates that chelate zinc and prevent absorption, and a low intake of meat (Reinhold 1971). Severe zinc deficiency is rare in humans, but marginal deficiency is not uncommon. It is found independent of socioeconomic status and can also occur in those who believe they are eating healthy diets (Sherertz and Goldsmith 1991). Severe deficiency causes loss of appetite and sense of taste, dermatitis, hair loss, diarrhea, neuropsychological disorders, retardation of growth, and disturbance in male sexual development and reproductive ability. Moreover, delayed wound healing and higher susceptibility to infections as a consequence of an impaired immune system may occur (D-A-CH 2008; Wolffram 1996; Sherertz and Goldsmith 1991).

When it comes to marginal deficiency, findings vary with the degree of deficiency. Increased susceptibility to infections, delayed wound healing, and aggravation of existing skin disorders can appear. As the zinc concentration begins to fall, an acral dermatitis begins to develop, which starts off as dry scaly, eczematous plaques in perioral, perineal, and acral areas. This plaque progresses to pustules and finally leads to erosive lesions (Neldner 1984; Neldner 1992). On the face, lesions occur around the eyes, nose, ears, cheeks, and neck; and the perineal rash can extend onto the lower abdomen and thighs. The rash is characteristically symmetrical and is accompanied by progressive alopecia, including loss of eyelashes in some cases

(Neldner 1993). Felons can appear on the hands and feet and erythema-like dermatitis on the palms (Ryan and Goldsmith 1996). Rutted nails (so-called Beau's lines) can be observed and attributed to zinc deficiency (Gaveau et al. 1987).

Almost all cases of zinc deficiency can be cured after 2–3 weeks of taking oral zinc supplements (Neldner 1984).

7.2.3 Impact on the Skin

Many skin disorders have been associated and treated with zinc (e.g., dandruff, acne, diaper rash), and oral zinc therapy has been shown to be effective in treating inflammatory conditions (Schwartz et al. 2005). Success rates for treatment with zinc vary depending on the disease, mode of application, and zinc salt used (Nitzan and Cohen 2006).

7.2.3.1 Zinc and Skin Appearance

Zinc has been shown to protect the skin cell membrane. A negative correlation was observed between zinc intake and skin evenness as well as the hydration of facial areas. However, as it is thought that the activity of the antioxidant enzymes, rather than the serum mineral concentration, has a positive influence on the moisture content of the facial areas excessive zinc intake is not recommended for skin moisturizing (Bae et al. 2010).

7.2.3.2 Zinc in the Healing Wound

For more than 3,000 years, orally supplemented or topically applied zinc in the form of zinc oxide or calamine has been used in the treatment of skin wounds. Admittedly, it is still unclear how much additional zinc a wound needs for the healing process to be enhanced or to what extent topical applied zinc is absorbed. After more than 100 published reports, the precise role of supplementary zinc therapy remains unclear (Lansdown 1996), but observations show that the requirement for zinc in the skin is higher during the process of healing (Henzel et al. 1970; Schwartz et al. 2005); and in individuals with dietary zinc deficiency or hereditary hypozincemia, zinc therapy is indicated for wound healing (Lansdown 1996).

Not only is zinc a constituent of enzymes that play a central role in reconstruction of a wound's matrix (Henzel et al. 1970), it also plays a significant role in the extracellular matrix, cell migration, protein synthesis, and modulation of inflammatory cytokines such as decreasing production of tumor necrosis factor-α (TNFα) (Bowe and Shalita 2008). In addition, zinc has an effect on the expression of integrins, which are cell surface proteins that mediate interactions between the cell and the various extracellular matrix proteins surrounding it. The effect of zinc was particularly notable on these integrins, affecting cellular mobility during the proliferative phase of wound healing (Schwartz et al. 2005).

7.2.3.3 Zinc and Acne

As patients with acne were reported to have low levels of serum zinc (Fitzherbert 1977), acne itself seems to be a symptom of zinc deficiency. It may be due to overproduction of male hormones, which was observed in cases of zinc deficiency. Used both internally and topically, zinc works to clear the skin by reducing oil production and may be effective in controlling the formation of acne lesions or help those already present to heal quicker by its antiinflammatory effect (Bowe and Shalita 2008).

During the 1970s, evidence was provided that acne improved with oral zinc supplementation in zinc-deficient patients (Fitzherbert 1977; Michaëlsson et al. 1977). Subsequently, trials showed that orally taken zinc was effective in the treatment of severe and inflammatory acne—more so than mild or moderate acne. Although these trials did not control for other dietary factors, oral zinc salt supplementation has been shown to be equal to or less effective than oral tetracyclines. However, the oral doses of zinc used in most of these studies (zinc gluconate 200 mg/day, zinc sulfate 400 or 600 mg/day) were associated with nausea, vomiting, abdominal cramps, and diarrhea (Bowe and Shalita 2008; Fitzherbert 1977; Michaëlsson et al. 1977; Rebello et al. 1986).

7.2.3.4 Zinc and Psoriasis

To which extent zinc is involved in the pathogenesis of psoriasis is not yet clear, but it seems that oral zinc supplementation has no effect on the skin manifestation of psoriasis (Burrows et al. 1994), although it ameliorates conditions of psoriatic arthropathy (Frigo et al. 1989; Clemmensen et al. 1980).

7.2.3.5 Zinc and Hair Loss

There is a widespread belief that zinc deficiency can cause hair loss in humans, which is seen only with severe zinc deficiency (Sherertz and Goldsmith 1991). However, there is no evidence of a link between low serum zinc to chronic telogen effluvium (CTE) (Arnaud et al. 1995) or alopecia areata (Ead 1981).

7.2.3.6 Zinc and Herpes Simplex Infections

Some studies indicate that oral zinc treatment may be considered as part of overall immune-enhancing therapy for patients with recurrent herpes simplex infections, leading to a reduction in attack frequency as well as a reduction in the duration and severity of outbreaks (Gaby 2006).

7.2.3.7 Topically Applied Zinc

Diaper Dermatitis

Zinc oxide, often combined with castor oil, has a special place in the treatment of diaper dermatitis (diaper rash). Although the ointments are often thought to be primarily a barrier cream separating the skin from the irritating excreted matter, it is likely that the antiinflammatory and antimicrobial activity of zinc oxide can help alleviate the symptoms (Schwartz et al. 2005; Scheinfeld et al. 2006).

Dandruff

Zinc pyrithione is an antifungal agent that is effective as an antidandruff agent and for seborrheic dermatitis (Bissett 2009). Zinc itself may also have an impact beyond the antifungal action: The antiinflammatory effect may reduce hyperproliferation of the dandruff skin (Schwartz et al. 2005). Moreover, zinc inhibits the enzyme responsible for converting testosterone to its biologically active metabolite, which mediates most of the actions typically ascribed to androgenic hormones, including stimulation of sebaceous gland activity (Sugimoto et al. 1995; Stamatiadis et al. 1988). Thus, reduction of sebum secretion is known to result from zinc exposure (Demetree et al. 1980) and could be an additional benefit of the use of zinc-containing antidandruff treatments (Schwartz et al. 2005).

Acne

With regard to acne, topically applied zinc has been proven to have positive effects due to its antimicrobial and antiinflammatory effects as well as reduction of hyperproliferation and high sebum production. Zinc is bacteriostatic against *Propionibacterium acnes*, one of the bacteria involved in causing acne. Zinc inhibits chemotaxis (Bowe and Shalita 2008) and may also directly inhibit bacterial lipases from converting sebum triglycerides to fatty acids, which are presumably more potent mediators of acne (Rebello et al. 1986). A combination of an antibiotic with a soluble zinc salt has resulted in decreased sebum production, as shown in clinical studies (Piérard-Franchimont et al. 1995). Comparison of the zinc-containing antibiotic treatment with one without zinc (although also at a lower antibiotic level) demonstrated the zinc-containing product to have greater lesion reduction efficacy (Habbema et al. 1989).

Herpes Simplex Skin Infections

Topically applied zinc preparations can shorten the duration of herpes simplex skin infections and possibly prevent recurrences. Although most studies used zinc sulfate, zinc monoglycerolate or zinc oxide-glycine may also be beneficial. Zinc oxide, however, seems not to be effective because it does not release a sufficient amount of zinc ions to exert an antiviral effect (Gaby 2006).

7.3 Iron

Iron is the most abundant trace element in the body. The adult human body contains approximately 2–5 g of iron, about 60% of which is bound to hemoglobin, 25% to ferritin and hemosiderin, and about 15% to myoglobin and enzymes (Anke et al. 1999).

7.3.1 Functions

- Transport of oxygen in the form of hemoglobin in erythrocytes
- Storage of oxygen as myoglobin in muscle cells
- Generation of energy in mitochondrial cytochromes
- Constituent in groups of enzymes (e.g., cytochrome P450 system)
- Production of neurotransmitters in the brain
- Factor in collagen synthesis
- Hydroxylation of proline and lysine (Table 7.3)

Good sources of iron are meat, liver, soy flour, millet, lentils, and green vegetables. The bioavailability of iron from food varies from <2% in high-fiber vegetables to 15–20% in meat. Ascorbic acid, cysteine, methionine, and citrate enhance iron availability. Phytin, phosphate, phospholipids, and tannic acid build complexes with iron that inhibit the absorption (D-A-CH 2008).

7.3.2 Deficiency

Iron deficiency represents a public health problem; its frequency ranges from <10% to as high as 70% among various ethnic and socioeconomic groups.

Table 7.3 Recommended daily intake for iron for adults

Healthy people/condition	RDA (mg)[a]
Men	8–10
Women	15–18
Pregnancy	27–30
Lactation	9–20

[a]Values vary depending on the source in the literature (Food and Nutrition Board, Institute of Medicine, National Academy of Sciences 2001; D-A-CH 2008)

Symptoms caused by iron deficiency include fatigue, lack of energy, loss of appetite, and increased susceptibility to infections (D-A-CH 2008). From a dermatological perspective, the iron-deficiency state can lead to dry skin, brittle hair, hair loss, rutted fingernails (Beau's lines), inflammation of the oral mucosa and tongue, pruritus, chronically sustained inflammation, dermatitis herpetiformis, and photodermatitis as well as changes in the overall appearance such as pallor and blue sclerae (Sato 1991).

7.3.3 *Impact on the Skin*

7.3.3.1 Pruritus

One of the many causes of puritus is iron deficiency with or without accompanying anemia. A Finnish study revealed that men and women who have low serum hemoglobin and ferritin concentrations suffer significantly more often from pruritus than people with normal iron values (Takkunen 1978).

7.3.3.2 Iron and Hair

In the presence of iron deficiency, the hair appears lusterless, dry, brittle, and spliced. It may be due to impaired keratin production (Sato 1991). The major cause of hair loss in women before the age of 50 is poor nutrition, which affects about 30% of women. Increased and persistent hair shedding (chronic telogen effluvium, or CTE) as well as reduced hair volume are the principal changes that occur. The main cause appears to be depleted iron stores accompanied by suboptimal intake of the essential amino acid L-lysine (Rushton et al. 2002).

As early as the 1960s, the role of iron as an etiological factor in diffuse hair loss in nonanemic women with an iron deficiency was demonstrated (Hård 1963). More recent studies succeeded in a significant decrease in hair shedding as well as in the percentage of hair in the telogen phase in women with CTE after supplementation with 72 mg iron and 1.5 g L-lysine daily for 6 months. In these studies, a significant increase in the mean serum ferritin concentration was found in the treated groups. This implies that there is a minimum ferritin concentration required to optimize treatment in those with diffuse hair loss. Although the exact level of serum ferritin that should be reached is still not fully resolved, 70 μg/L seems to be the level to achieve. The amount of supplementation or dietary change needed to maintain this serum ferritin concentration is the subject of continuing research. According to a preliminary analysis of data, a daily iron supplement of 24–48 mg seems to be needed by menstruating women, with a lower level required in postmenopausal women (Rushton et al. 2002).

7.3.3.3 Iron and Nails

Brittle and rutted fingernails (Beau's lines) are often observed with moderate iron deficiency. In advanced cases, the nails have an even, plain, spoon-shaped, convex structure, predominantly, but not exclusively, the index and middle fingers. As these characteristically spoon-shaped nails are not seen in other conditions, this is an infallible indication for iron deficiency. Recovery takes a long time after introducing therapy (Sato 1991; Scheinfeld et al. 2007).

7.4 Copper

As an essential trace element, copper is, with 50–150 mg, third in abundance in the human body after iron and zinc (Anke et al. 1999).

7.4.1 Functions

- Compartment of several vital metalloenzymes that are involved, for example, in blood clotting, detoxification of free radicals, and oxidation–reduction
- Involvement in iron metabolism
- Constituent of tyrosinase, which functions in melanin production
- Essential for the production of mature collagen and elastin as well as hair maturation (Table 7.4)

Good sources of copper are cereal products, fish, liver, eggs, cacao, chocolate, coffee, nuts, sunflower seeds, lentils, peas, and some other green vegetables. The bioavailability of copper varies between 10% and 70% (D-A-CH 2008).

Several dietary interactions of copper with other nutrients decrease or inhibit copper absorption. Especially, sucrose or fructose, proteins, amino acids, iron, high-dose ascorbic acid supplements, high intake of calcium and/or phosphorus, and high levels of dietary zinc have been shown to adversely influence copper absorption and bioavailability (Scientific Committee on Food 2003).

Table 7.4 Recommended daily intake for copper for adults

Healthy people/condition	RDA (mg)[a]
Men	0.9–1.5
Women	0.9–1.5
Pregnancy	1.0
Lactation	1.3

[a]Values vary depending on the source in the literature (Food and Nutrition Board, Institute of Medicine, National Academy of Sciences 2001; D-A-CH 2008)

Copper deficiency is rare, and taking copper supplements can be dangerous as the dose for intoxication is relatively low. A tolerable upper intake level of 5 mg/day was derived; but this upper level is not applicable during pregnancy or lactation because of inadequate data relating to this critical life stage (Scientific Committee on Food 2003). Copper intoxication causes vomiting, abdominal pain, diarrhea, and yellow discoloration of the skin. Uptake of more than 10 g causes severe damage to the liver and kidneys as the mineral concentrates in these organs (Wolffram 1996).

7.4.2 *Deficiency*

Copper deficiency can result from long-term parenteral nutrition or chronic malabsorption, although high-dosage supplementation of iron or zinc can lead to copper deficiency. Moreover, continuous use of antacids, chronic diarrhea, and intestinal inflammation have a negative effect on copper absorption.

The most notable occurrence in those with copper deficiency is depigmentation of hair and skin, and structural changes in hair occur as well. To an extreme they are observed in patients with Menkes' kinky hair syndrome, an X-chromosome-linked disorder of copper metabolism. The overall appearance of the hair is lusterless, like straw, and depigmented; microscopically, the hair twists around its axis. Other cutaneous manifestations of that disease include deficient eyebrows and eyelashes, follicular hyperkeratosis, soft inelastic skin, and depigmentation. Cerebral degeneration and severe arterial disease are the most important systemic manifestations and usually lead to death at 3–4 years of age (Ryan and Goldsmith 1996; Madsen and Gitlin 2007).

Slight copper deficiency leads to anemia with leukopenia and neutropenia, increased sensitivity to oxidative damage, hypercholesterolemia, hypertriglyceridemia, glucose intolerance, impaired immune reaction, poor wound healing, osteoporosis, faintness, and fatigue.

Structural damage in vascular walls due to defective formation of elastin has also been noted and can be associated with an increased risk for aneurysm. Although one might expect changes in skin elasticity given the role of copper in elastin formation, it is not generally observed (Ryan and Goldsmith 1996; Madsen and Gitlin 2007).

7.4.2.1 Impact on the Skin

Topical use of copper–peptide complex can provide facial skin antiaging effects (Burke 2006).

Copper-containing creams have been found to firm the skin and help restore some elasticity. Compared to a popular skin care treatment and a placebo, a cream containing copper peptides demonstrated rapid, visual overall improvement in skin roughness, clarity, fine lines, wrinkles, and overall photodamage (Leyden et al. 2002a, b).

Immediate and lasting improvement of skin appearance after application of a bimineral complex of elementary zinc and copper, which had shown antiinflammatory activity and up-regulation of collagen and elastin gene expression in vitro, was observed in clinical studies. Evaluations via clinical grading of efficacy parameters, high-resolution digital imaging, and subject self-assessments revealed statistically significant and visible improvement of overall appearance, skin radiance, under-eye bags, under-eye wrinkles, under-eye dark circles, sagging, cheek wrinkles, pigmentation, radiance, fine lines, as well as global lifting and firming when the preparation was applied at least once a day (Chantalat et al. 2010a, b; Lanctin et al. 2010).

7.5 Selenium

The total body pool of selenium has been estimated to be 5–15 mg in adults (Patterson et al. 1989).

7.5.1 Functions

- Antioxidant as a constituent of glutathione peroxidase
- Part of an antioxidant system that protects cell membranes and structural membranes from lipid peroxidation
- Immunomodulation via regulation of immunoglobulin G (IgG) production and stimulation of leukocyte activity
- Regulation of thyroid hormone metabolism
- Needed to maintain tissue elasticity (Table 7.5)

Good sources of selenium are meat, fish, seafood, eggs, lentils, asparagus, soy beans, Brazil nuts, and whole grain cereals (D-A-CH 2008). People who do not follow a balanced diet (e.g., strictly vegan) or energy- and protein-reduced diets are prone to selenium deficiency.

Table 7.5 Recommended daily intake for selenium for adults

Healthy people/condition	RDA (μg)[a]
Men	30–70
Women	30–70
Pregnancy	30–70
Lactation	30–70

[a]Values vary depending on the source in the literature (D-A-CH 2008; Food and Nutrition Board, Institute of Medicine, National Academy of Sciences 2000)

The balance between selenium deficiency and toxicity can be fragile. The therapeutic window for supplementation is narrow and not without risk. Because of the increase in self-medication with selenium, several cases of acute selenium intoxication occurred because of falsely specified preparations (Wolffram 1996).

Intake of 250 mg selenium as a single dose or multiple doses of 27–31 mg resulted in acute toxicity with nausea, vomiting, nail changes, dryness of hair, hair loss, tenderness and swelling of fingertips, fatigue, irritability, and garlicky breath (Jensen et al. 1984; World Health Organization 1987).

A tolerable upper intake level of 300 μg/day was derived for adults. This value covers selenium intake from all sources of food, including supplements (Scientific Committee on Food 2000).

7.5.2 Deficiencies

No specific selenium deficiency condition has been described in humans. Obvious deficiency symptoms occur in Keshan disease, which is related to a combination of coxsackie virus and low selenium uptake, causing inflammation of the heart muscle, leading to cardiomyopathy. Manifestations are recognized in skin and hair in the form of depigmentation, which is also observed with long-term deficiency due to long-term parenteral nutrition (Alexander 2007).

7.5.3 Impact on the Skin

7.5.3.1 Selenium in Acne

Because the selenium-dependent glutathione peroxidase enzyme activity is low in acne patients, it has been theorized that selenium would be of value. Indeed, low levels of blood selenium have been documented in acne patients. A study examined the effect of supplementation with selenium (400 μg) and tocopherol (20 mg) daily for 12 weeks. The combination led to reduced severity of persistent acne and reduced scarring, especially in those with low baseline glutathione peroxidase activity (Michaëlsson and Edqvist 1984; Michaëlsson 1990).

7.5.3.2 Selenium and Psoriasis

Observations showed low selenium concentration and depressed glutathione peroxidase activity in patients with psoriasis, especially in those who had been suffering from the disease for 3 years or more (Serwin et al. 2003). Inflammatory reactions in the skin may be the reason for selenium deficiency. A study in psoriatic patients showed alleviation of the clinical manifestations after drinking and immersion in inorganic selenium-rich spa water (selenate 70 μg/L) for 3 weeks. Patients who responded to treatment had a significant increase in their plasma selenium level (Pinton et al. 1995). However, supplementation with inorganic selenomethionine did not lead to clinical improvement (Serwin et al. 2003).

7.5.3.3 Topical Application

Selenium sulfide is an antifungal agent that is effective as an antidandruff agent (Bissett 2009). Topical application of selenomethionine prevents UV-induced erythema and is discussed as providing facial antiwrinkle effects (Burke 2006).

7.6 Chromium

Chromium is ubiquitous. Chromium ions are constituents of the glucose tolerance factor that is important for protein, glucose, and lipid metabolism (Anke et al. 1999).

7.6.1 Functions

- Responsible for production of certain enzymes
- Influences carbohydrate, lipid, and protein metabolism
- Effect on insulin action, part of the "glucose tolerance factor"
- Helps to fight acne and reduce infections

Currently, there is no formal RDA for chromium. The U.S. Food and Nutrition Board (Food and Nutrition Board, Institute of Medicine, National Academy of Sciences 2001) and the Societies for Nutrition of Germany (DGE), Austria (ÖGE), and Switzerland (SGE) (D-A-CH 2008) derived adequate intakes for chromium (Table 7.6). The World Health Organization considered that long-term supplementation of chromium should not exceed 250 μg/day (World Health Organization 1996).

Table 7.6 Estimated adequate intake of chromium for adults

Healthy people	Estimated adequate intake (μg/day)[a]
Men	30–100
Women	25–100

[a]Values vary depending on the source in the literature (Food and Nutrition Board, Institute of Medicine, National Academy of Sciences 2001; D-A-CH 2008)

Good sources of chromium are meat, liver, eggs, seafood, whole grain products, oat flakes, lettuce, green beans, broccoli, tomatoes, potatoes, prunes, nuts, cacao, and mushrooms. Chromium in food is not easily absorbed. Some research has indicated that almost 90% of the population could be deficient. The mineral is lost when food is processed. A diet that is high in sugar intake can also cause a deficiency of chromium (D-A-CH 2008).

7.6.2 Impact on the Skin

Chromium has been shown to reduce skin infection rates and thus may help decrease the symptoms of acne. Moreover, it is estimated that 90% of people suffering from acne also exhibit excessive or unstable blood glucose levels. The human body recognizes the presence of insulin very quickly, and chromium takes part in insulin metabolism; thus, a sufficient amount of chromium in the body helps maintain a constant glucose level. Consequently, trials showed that supplementation with 400 μg chromium daily improved skin condition in patients with severe acne (McCarty 1984).

7.7 Silicon

Silicon is present in biological material as a silanate, an ether (or ester-like) derivative of silicic acid, which may play a role in the structure of glycosaminoglycans and their protein complexes. Despite the fact that silicon is involved in important processes, there is little known about its mode of action or physiological requirements (Nielsen et al. 2003).

7.7.1 Functions

- Along with calcium, involved with growth and maintenance of strong bones
- Formation and structure element of connective tissue, ligaments, and tendons
- Involvement in cell metabolism and division
- Growth of hair, skin, and fingernails
- Favors biological moisturization of the dermis and epidermis because of the ability to maintain water bound in tissues

A form of silicon is a home remedy for problems with weakening bones, painful joints, and aging skin, although there is no clear scientific evidence for actual improvement in patients with these conditions (Paillet 2000).

7.7.2 *Dietary Recommendation*

Even though measurable responses of humans to variations in dietary silicon intake have been demonstrated, no estimated average requirement or adequate intake has been set because the data available are still insufficient (Food and Nutrition Board, Institute of Medicine, National Academy of Sciences 2001). Probably, a daily intake of 5–10 mg of highly absorbable silicon will be found adequate (Nielsen et al. 2003), but not all food silicon is highly absorbable.

Good sources of silicon are brown rice, cereal products, leafy green vegetables, root vegetables, seafood, and soybeans. Because it is made from grains, beer also is a dietary source of silicon (D-A-CH 2008).

7.7.3 *Deficiency*

Signs of silicon deficiency have not been described for humans. Nonetheless, the nature of silicon deficiency in animals has led to the speculation that the lack of silicon is involved in several human disorders, including atherosclerosis, osteoarthritis, and hypertension as well as the aging process (Nielsen et al. 2003).

7.7.4 *Impact on the Skin*

Orthosilicic acid, a bioavailable form of silicon that has been shown to stimulate collagen synthesis in skin fibroblasts, showed a significant positive effect on the skin surface (skin roughness) and mechanical properties as well as on hair and nail brittleness in women with photodamaged facial skin after supplementation with 10 mg daily for 20 weeks (Barel et al. 2005).

Take Home Messages

- Several trace elements, including zinc, iron, copper, selenium, chromium, and silicon, affect the condition of the skin.
- It is especially deficiencies in these minerals, which can manifest in skin, hair, or nails. In most cases these phenomena can be corrected by supplementation of the respective mineral.
- Moreover, some minerals, orally taken or topically applied, have shown to be effective in the treatment of skin diseases or can help preserve a youthful appearance of the skin.
- Supplements should be used carefully. Therapeutic doses should be taken only under medical supervision as excessive intake of minerals may lead to intoxication.

References

Alexander J. Selenium. Novartis Found Symp. 2007;282:143–9. discussion 149-53, 212–218.

Anke M, Schürmann K. Spurenelemente (German). In: Biesalski HK, Fürst P, Kasper H, Kluthe R, Pölert W, Puchstein C, Stähelin HB, editors. Ernährungsmedizin. Thieme Verlag, Stuttgart; 1999. p. 173–186.

Arnaud J, Beani JC, Favier AE, Amblard P. Zinc status in patients with telogen defluvium. Acta Derm Venereol. 1995;75(3):248–9.

Bae HS, Choi SI, Ahn HS. Nutritional and antioxidant status by skin types among female adults. Nutr Res Pract. 2010;4(3):215–21. Epub 2010 Jun 28.

Barel A, Timchenko A, Van den Berghe D, Clarys P. Effect of oral intake of choline-stabilized orthosilicic acid on skin, nails, and hair in women with photodamaged facial skin. American Academy of Dermatology, 63th Annual Meeting, P208, New Orleans, 2005.

Bissett DL. Common cosmeceuticals. Clin Dermatol. 2009;27(5):435–45.

Bowe WP, Shalita AR. Effective over the counter acne treatments. Semin Cutan Med Surg. 2008;27:170–6.

Burke KE. Topical nutritional antioxidants. In: Draelos ZD, Thaman LA, editors. Cosmetic formulation of skin care products. New York: Taylor & Francis; 2006. p. 377–402.

Burrows NP, Turnbull AJ, Punchard NA, Thompson RP, Jones RR. A trial of oral zinc supplementation in psoriasis. Cutis. 1994;54(2):117–8.

Chantalat J, Bruning E, Liu JC, Sun Y. P1624 Biomimetic signaling technology generated by a bimineral complex reduces signs of photoaging in the eye area. American Academy of Dermatology, 68th Annual Meeting, Miami Beach, 2010a.

Chantalat J, Bruning E, Liu JC, Sun Y. P304 Biomimetic signaling technology generated by a bimineral complex shows efficacy in reducing clinical photoaging in a 12-week placebo controlled study. American Academy of Dermatology, 68th Annual Meeting, Miami Beach, 2010b.

Clemmensen OJ, Siggaard-Andersen J, Worm AM, Stahl D, Frost F, Bloch I. Psoriatic arthritis treated with oral zinc sulphate. Br J Dermatol. 1980;103(4):411–5.

D-A-CH: Deutsche Gesellschaft für Ernährung (DGE), Österreichische Gesellschaft für Ernährung (ÖGE), Schweizerische Gesellschaft für Ernährungsforschung (SGE), Schweizerische Vereinigung für Ernährung (SVE) (editors): Referenzwerte für die Nährstoffzufuhr [German]. Neuer Umschau Buchverlag, Neustadt an der Weinstraße, corrected reprinting of the 1. Edition, 2008.

Demetree JW, Safer LF, Artis WM. The effect of zinc on sebum secretion rate. Acta Derm Venereol. 1980;60:166–9.

Ead RD. Oral zinc sulphate an alopecia areata – a double blind study. Br J Dermatol. 1981;104(4):483.

Fitzherbert JC. Zinc deficiency in acne vulgaris. Med J Aust. 1977;2:685–6.

Food and Nutrition Board, Institute of Medicine, National Academy of Sciences. Dietary reference intakes for vitamin C, vitamin E, selenium and carotenoids. Washington, DC: National Academy Press; 2000.

Food and Nutrition Board, Institute of Medicine, National Academy of Sciences: Dietary Reference Intakes: Vitamin A, Vitamin K, Arsenic, Boron, Chromium, Copper, Iodine, Iron, Manganese, Molybdenum, Nickel, Silicon, Vanadium and Zinc. National Academy Press; 2001

Frigo A, Tambalo C, Bambara LM, Biasi D, Marrella M, Milanino R, et al. Zinc sulfate in the treatment of psoriatic arthritis (Article in Italian). Recenti Prog Med. 1989;80(11):577–81.

Gaby AR. Natural remedies for Herpes simplex. Altern Med Rev. 2006;11(2):93–101.

Gaveau D, Piette F, Cortot A, Dumur V, Bergoend H. Cutaneous manifestations of zinc deficiency in ethylic cirrhosis (Article in French). Ann Dermatol Venereol. 1987;114(1):39–53.

Habbema L, Koopmans B, Menke HE, Doornweerd S, De Boulle K. A 4% erythromycin and zinc combination (Zineryt) versus 2% erythromycin (Eryderm) in acne vulgaris: a randomized, double-blind comparative study. Br J Dermatol. 1989;121:497–502.

Hård S. Non- anemic iron deficiency as an etiological factor in diffuse loss of hair of the scalp of women. Acta Derm Venereol. 1963;43:562–9.

Henzel JH, DeWeese MS, Lichti EL. Zinc concentration within the healing wounds. Arch Surg. 1970;100:349–57.

Jensen R, Clossen W, Rothenberg R. Selenium intoxication. New York Morb Mortal Wkly Rep. 1984;33:157–8.

King JC, Shames DM, Woodhouse LR. Zinc homeostasis in humans. J Nutr. 2000;130:1360S–6.

Lanctin M, Nkengne A, Bertin C, Roure R. P1623A cosmetic product containing biomimetic signaling technology to improve the aging signs in the eye area. American Academy of Dermatology, 68th Annual Meeting, Miami Beach, 2010.

Lansdown AB. Zinc in the healing wound. Lancet. 1996;347:706–7.

Leyden JJ, Barkovic S, Appa Y. P67 the effect of tripeptide to copper ratio in two copper peptide creams on photoaged facial skin. American Academy of Dermatology 60th Annual Meeting, New Orleans, 2002a.

Leyden JJ, Stephens T, Finkey MB, Barkovic S. P68 skin care benefits of copper peptide containing facial cream. American Academy of Dermatology, 60th Annual Meeting, New Orleans, 2002b.

Madsen E, Gitlin JD. Copper deficiency. Curr Opin Gastroenterol. 2007;23(2):187–92.

McCarty M. High-chromium yeast for acne? Med Hypotheses. 1984;14:307–10.

Michaëlsson G. Decreased concentration of selenium in whole blood and plasma in acne vulgaris. Acta Derm Venereol. 1990;70(1):92.

Michaëlsson G, Edqvist LE. Erythrocyte glutathione peroxidase activity in acne vulgaris and the effect of selenium and vitamin E treatment. Acta Derm Venereol. 1984;64:9–14.

Michaëlsson G, Juhlin L, Ljunghall K. A double-blind study of the effect of zinc and oxytetracycline in acne vulgaris. Br J Dermatol. 1977;97:561–6.

Neldner KH. Nutrition, aging and the skin. Geriatrics. 1984;39(69–82):87–8.

Neldner KH. Vitamins, trace elements, and the skin. In: Moschella S, editor. Dermatology. 3rd ed. Philadelphia: WB Saunders; 1992. p. 1605–28.

Neldner KH. Acrodermatitis enteropathica and other zinc-deficiency disorders. In: Fitzpatrick TB, Eisen AZ, Wolff K, et al., editors. Dermatology in general medicine. 4th ed. New York: McGraw-Hill; 1993. p. 1826–33.

Nielsen FH, Hunt JR. Trace elements emerging as important in human nutrition; 2003 http://www.nutrientdataconf.org/PastConf/NDBC14/5-1_Nielsen.pdf

Nitzan YB, Cohen AD. Zinc in skin pathology and care. J Dermatol Treat. 2006;17(4):205–10.

Paillet C. Application of silicium for cosmetics. Skin and silicium. Fragr J. 2000;28(11):23–6.

Patterson BH, Levander OA, Helzlsouer K, McAdam PA, Lewis SA, Taylor PR, et al. Human selenite metabolism: a kinetic model. Am J Physiol. 1989;257:R556–67.

Piérard-Franchimont C, Goffin V, Visser JN, Jacoby H, Piérard GE. A double-blind controlled evaluation of the sebosuppressive activity of topical erythromycin-zinc complex. Eur J Clin Pharmacol. 1995;49(1–2):57–60.

Pinton J, Friden H, Kettaneh-Wold N, Wold S. Clinical and biological effects of balneotherapy with selenium-rich spa water in patients with psoriasis vulgaris. Br J Dermatol. 1995;133:344–7.

Rebello T, Atherton DJ, Holden C. The effect of oral zinc administration on sebum free fatty acids in acne vulgaris. Acta Derm Venereol. 1986;66:305–10.

Reinhold JG. High phytate content of rural Iranian bread: a possible cause of human zinc deficiency. Am J Clin Nutr. 1971;24(10):1204–6.

Rushton DH, Norris MJ, Dover R, Busuttil N. Causes of hair loss and the developments in hair rejuventation. Int J Cosmet Sci. 2002;24:17–23.

Ryan AS, Goldsmith LA. Nutrition and the skin. Clin Dermatol. 1996;14(4):389–406.

Sato S. Iron deficiency: structural and microchemical changes in hair, nails, and skin. Semin Dermatol. 1991;10(4):313–9.

Scheinfeld N. P2001 Efficacy of zinc oxide and petrolatum moisture barriers in the treatment of diaper dermatitis (DD) and DD complicated by candidiasis (DDCC) American Academy of Dermatology, 64th Annual Meeting, 2006.

Scheinfeld N, Dahdah MJ, Scher R. Vitamins and minerals: their role in nail health and disease. J Drugs Dermatol. 2007;6(8):782–7.

Schwartz JR, Marsh RG, Draelos ZD. Zinc and skin health: overview of physiology and pharmacology. Dermatol Surg. 2005;31(7 Pt 2):837–47.

Scientific Committee on Food: Opinion on "Opinion of the Scientific Committee on Food on the Tolerable Upper Intake Level of Copper" (expressed on 5 March 2003); http://ec.europa.eu/food/fs/sc/scf/out176_en.pdf

Scientific Committee on Food: Opinion on "Opinion of the Scientific Committee on Food on the Tolerable Upper Intake Level of Selenium" (expressed on 19 October 2000); http://ec.europa.eu/food/fs/sc/scf/out80g_en.pdf

Serwin AB, Wasowicz W, Gromadzinska J, Chodynicka B. Selenium status in psoriasis and its relations to the duration and severity of the disease. Nutrition. 2003;19(4):301–4.

Sherertz EF, Goldsmith LA. Nutritional influences on the skin; Neldner KH. The biochemistry and physiology of zinc metabolism; Danks DM. Copper deficiency and the skin. In: Goldsmith LA, editor. Physiology, biochemistry, and molecular biology of the skin. 2nd ed. New York: Oxford University Press; 1991. p. 1315–61.

Stamatiadis D, Bulteau-Portois MC, Mowszowicz L. Inhibition of 5 -reductase activity in human skin by zinc and azelaic acid. Br J Dermatol. 1988;119:627–32.

Sugimoto Y, López-Solache I, Labrie F, Luu-The V. Cations inhibit specifically type I 5 -reductase found in human skin. J Invest Dermatol. 1995;104:775–8.

Takkunen H. Iron-deficiency pruritus. JAMA. 1978;239(14):1394.

Wolffram S. Spurenelemente (German). In: Forth W, Henschler D, Rummel W, Starke K, editors. Allgemeine und spezielle Pharmakologie und Toxikologie. Heidelberg/Berlin/Oxford: Spektrum Akademischer Verlag; 1996. p. 657–63.

World Health Organization. Selenium, Environmental Health Criteria 58. Geneva: World Health Organization; 1987.

World Health Organization. Trace elements in human nutrition and health, (A Report of a re-evaluation of the role of trace elements in human health and nutrition). Geneva: World Health Organization; 1996.

Chapter 8
Probiotics and Skin

Robert J. Boyle, Sampo J. Lahtinen, and Mimi L.K. Tang

Core Messages

- Our cutaneous and intestinal microbiota have a close relationship with skin health, so manipulation of these microbiota may have health benefits for the skin.
- There is little direct evidence that topical or oral probiotics administered beyond the early perinatal period lead to long-term modification of the intestinal or cutaneous microbiota composition after treatment has ceased.
- Nevertheless, there are promising data suggesting that probiotics administered by either route may have significant beneficial effects on the skin.
- Orally ingested probiotic bacteria have been widely investigated for a potential role in treating or preventing the common skin disease eczema.
- The weight of evidence suggests that some probiotic bacteria administered both prenatally and postnatally may be effective for preventing eczema, perhaps through promotion of healthy immune development.
- Current evidence suggests that orally administered probiotic bacteria are not effective for the treatment of established eczema.

8.1 Introduction

Approximately 90% of cells in each human are constituents of the large intestinal microbiota, so we are in at least one sense more microbe than human. This intense colonization is no evolutionary accident but provides survival advantages for both microbes and humans. The intestinal microbiota resides in close communication

R.J. Boyle (✉)
Department of Paediatrics, Imperial College London,
Wright Fleming Building, Norfolk Place, London W2 1PG, UK
e-mail: r.boyle@imperial.ac.uk

A. Pappas (ed.), *Nutrition and Skin: Lessons for Anti-Aging, Beauty and Healthy Skin*,
DOI 10.1007/978-1-4419-7967-4_8, © Springer Science+Business Media, LLC 2011

with the gut-associated lymphoid tissue, our largest immune organ, and an important role of the intestinal microbiota is to promote healthy immune development. Deviation from healthy intestinal microbiota has been associated with the development of various immune disorders, including inflammatory skin disorders such as eczema (Bjorksten et al. 2001). Thus, in principle, manipulation of the intestinal microbiota might be an effective way to influence cutaneous immune-mediated diseases, particularly those that are associated with alterations in intestinal function and/or microbiota composition.

The use of probiotic bacteria to alter the intestinal microenvironment and therefore impact on immune function is one approach to the management of skin disease. Probiotic preparations have largely been administered via the oral route, but more recently topical use has been investigated. The latter approach derives from the importance of our cutaneous microbiota—we typically have 10^4–10^6 microbes/cm^2 on our skin, including bacteria and fungi—and aims to influence immune cells or epithelial cells in the skin.

Although the composition of the skin microbiota in an individual is less stable than the intestinal microbiota, it varies significantly between healthy and unhealthy skin, even within the same individual (Gao et al. 2008). For example, the skin follicles of individuals with acne contain increased numbers of organisms such as *Propionibacterium acne* (Bek-Thomsen et al. 2008; Dekio et al. 2007). Abnormalities of the skin microbiota also have been well documented in psoriasis and eczema (Gao et al. 2008). It is unclear to what extent these changes in microbiota lead to or exacerbate disease or if they are simply a consequence of an altered cutaneous microenvironment secondary to the disease process. With these principles in mind, we spend the major part of this chapter discussing the empirical evidence whether topical and orally ingested probiotics have a role in the treatment or prevention of skin disease.

8.2 Probiotics

Beneficial microbes in the form of *Streptococcus thermophilus* and *Lactobacillus bulgaricus* in fermented milk have been ingested by humans for thousands of years in the belief that fermented products have health benefits. During the early twentieth century, the Russian immunologist Elie Metchnikoff proposed that lactic acid bacilli may have beneficial health effects and attributed his own longevity to regular ingestion of beneficial microbes (Metchnikoff 1906). The proposed health benefits of probiotics have undergone increasingly rigorous scientific evaluation in recent years, and there is now strong evidence for their use in treating and preventing a number of human intestinal diseases. However, community use of probiotics is much wider than these specific indications, and probiotics have become an important commercial commodity.

Probiotics are most commonly defined as "live microorganisms which when administered in adequate amounts confer a health benefit on the host" (FAO/WHO 2002). It is believed by many experts that the ideal probiotic should remain viable at the level of the intestine and should adhere to the intestinal epithelium to confer a significant health benefit. There is some evidence to support the importance of viability in human studies, with viable bacteria having greater immunological effects than nonviable bacteria (Kaila et al. 1995; Kirjavainen et al. 2003). Some of the best characterized probiotics have also been shown to adhere strongly to intestinal epithelium in both in vitro and in vivo studies (Alander et al. 1999). Probiotics must also be resistant to gastric acid digestion and to bile salts to reach the intestine intact, and they should be nonpathogenic. Most probiotics are strains of *Bifidobacterium* or *Lactobacillus* species. Some have been isolated from the intestinal microbiota of healthy humans; others have been isolated from fermented dairy products. Species from other bacterial genera such as *Streptococcus, Bacillus*, and *Enterococcus* have also been used as probiotics, but there are concerns surrounding the safety of some of these probiotics because they contain many pathogenic species, particularly within the genus *Enterococcus* (FAO/WHO 2002). Nonbacterial microorganisms such as yeasts from the genus *Saccharomyces* have also been used as probiotics for many years.

Probiotics have been advocated for the prevention and/or treatment of a diverse range of disorders, from acute gastroenteritis to intestinal neoplasia. The evidence for their efficacy in many such disorders is not strong, but there is well-established benefit in a small number of conditions. The strongest evidence for their use is in the management of diarrheal diseases. For example, a meta-analysis of randomized controlled trials (RCTs) has shown that many probiotics are effective in preventing antibiotic-associated diarrhea (D'Souza et al. 2002). They include the yeast *Saccharomyces boulardii,* the bacterium *Lactobacillus acidophilus* in combination with *L. bulgaricus, L. rhamnosus* strain GG (American Type Culture Collection [ATCC] 53103; LGG) and *Enterococcus faecium* strain SF68.

A separate meta-analysis of randomized controlled trials has shown a variety of probiotics (including *Lactobacillus* species, *Enterococcus* species, and *S. boulardii*) to be effective in the treatment of infective diarrhea in both adults and children (Allen et al. 2003). In this analysis, probiotics were found to reduce the mean duration of diarrhea by more than 30 h. Apart from acute gut infections, probiotics have also been evaluated in the treatment of irritable bowel syndrome (IBS) and various types of inflammatory bowel disease (IBD). Several studies on probiotics and IBS have been performed using several probiotic candidates (Aragon et al. 2010). Although the evidence for probiotics as a treatment for IBS is promising, more and larger clinical studies are needed to establish the benefits for each specific probiotic strain or strain combination. The clinical evidence for probiotics in the treatment of IBD is somewhat conflicting. Studies in Crohn's disease have often failed to show clinical benefit, whereas studies of other forms of IBD, such as ulcerative colitis or pouchitis, have demonstrated potential benefit (Gionchetti et al. 2000). Finally, there is good evidence that probiotics added to enteral feeds are able to prevent an intestinal complication of premature birth (necrotizing enterocolitis), at least in infants weighing >1,000 g at birth (Alfaleh and Bassler 2008).

Probiotics have been proposed to have benefits outside the gastrointestinal tract. They may be mediated, for example, by interactions between the host immune system and the microbes or may result from antipathogenic activities of probiotics. A recent focus of probiotic research has been the prevention of common winter infections by probiotics, and several recent studies have demonstrated potential benefits for probiotics, such as a combination of *L. acidophilus* NCFM and *B. lactis* Bi-07 (Leyer et al. 2009) or *L. casei* DN-114 001 (Guillemard et al. 2010). Probiotic therapy has also been explored in skin diseases, particularly for the treatment and prevention of eczema.

8.3 Eczema

There has been a rapid rise in the prevalence of allergic and autoimmune disorders in recent decades. Reduced exposure to microbial stimuli associated with a modern lifestyle is suggested to have contributed to this rise. The intestinal microbiota represents the greatest microbial exposure throughout life, and acquisition of the early intestinal microbiota is the newborn infant's first major microbial challenge. Intestinal microbiota development has been shown to play an important role in immune regulation and the induction and/or maintenance of tolerance to environmental and self antigens (Sudo et al. 1997). Infants with eczema have altered intestinal microbiota compared to nonallergic children, with reduced numbers of the genus *Bifidobacterium* and increased levels of clostridia (*Clostridium difficile*), staphylococci (*Staphylococcus aureus*), and *Escherichia coli* (Bjorksten et al. 2001; Kalliomaki et al. 2001a; Penders et al. 2007; Watanabe et al. 2003). These differences were observed prior to the onset of disease (Bjorksten et al. 2001; Kalliomaki et al. 2001a; Penders et al. 2007), and the presence of *E. coli* or *C. difficile* at age 1 month was associated with increased risk for developing eczema, recurrent wheeze, and allergic sensitization. Manipulation of the intestinal microbiota during infancy may therefore provide an approach to the prevention or treatment of allergic skin conditions. It is unclear whether probiotics are able to modulate early development of intestinal microbiota when administered early in life, but some studies support this possibility (Gueimonde et al. 2006; Lahtinen et al. 2009) and therefore suggest a mechanism through which their use might alter immune development and target allergic skin disorders in the developing human.

8.3.1 Probiotics for Treating Eczema

Probiotic bacteria have been quite widely investigated for effects on the treatment of eczema with or without associated food allergy, in infants and children. Most of the studies involving eczema have evaluated *Lactobacillus* species either alone or in combination with other probiotic bacteria. Early studies in small numbers of infants and children reported improvement in eczema (SCORAD or symptoms) following treatment with *Lactobacillus rhamnosus* GG (LGG), *Bifidobacterium lactis* Bb-12, or *B. breve* M-16 V. In two of these studies, probiotic treatment resulted in more rapid alleviation of the eczema than with placebo: SCORAD improved significantly in the probiotic treatment groups but not the placebo groups at early time points; however, by the 2- to 6-month follow-up the eczema was markedly diminished in both treatment groups, with similar SCORAD scores in the treatment and placebo groups (Isolauri et al. 2000; Majamaa and Isolauri 1997). A larger study of *L. fermentum* VR1-003PCC in children with eczema also demonstrated improvements in eczema severity and extent in probiotic-treated infants but not in placebo-treated infants. However, SCORAD, parental perception of eczema, the impact of eczema on the family, and topical corticosteroid use were not significantly different between the active and placebo groups (Weston et al. 2005).

In contrast to these studies, many others have failed to confirm a beneficial effect of probiotics for the treatment of eczema (Brouwer et al. 2006; Folster-Holst et al. 2006; Gruber et al. 2007; Rosenfeldt et al. 2003; Sistek et al. 2006; Viljanen et al. 2005a). A small crossover study assessing the efficacy of a probiotic yoghurt containing *L. paracasei* Lpc-37, *B. lactis* 420, and *L. acidophilus* 74–2 in 15 adults with atopic dermatitis reported a trend toward reduced SCORAD with probiotic treatment (15.5 point reduction); however, this effect did not reach statistical significance ($P = 0.081$) (Roessler et al. 2008). Interestingly, in two studies showing no overall effect, subgroup analyses revealed improvements in SCORAD following treatment with LGG (Viljanen et al. 2005a) or the combination of *L. rhamnosus* HN001 and *B. lactis* HN019 (Sistek et al. 2006) for a subgroup of children with immunoglobulin E (IgE)-associated ("atopic") eczema (Sistek et al. 2006; Viljanen et al. 2005a). Probiotic treatment was associated with low-grade inflammation, as evidenced by moderate increases in C-reactive protein (CRP) and interleukin-6 (IL-6), which the authors suggested may suppress inflammation by inhibiting production of other inflammatory factors and inducing a regulatory immune response (Viljanen et al. 2005b). There is no direct evidence, however, that probiotics induce regulatory immune responses or inhibit inflammatory responses when used to treat individuals with established eczema.

Our recent systematic review evaluating the use of probiotics for treating eczema (Boyle et al. 2008) concluded that probiotics do not appear to be an effective treatment, and there is insufficient evidence to support their use for this condition. Our Cochrane systematic review (Boyle et al. 2008) included 12 studies (summarized in Table 8.1) but not the recent study in adults by Roessler et al. (2008). We found no significant reduction in eczema symptoms with probiotic treatment compared with placebo—mean difference 0.90 points on a 20-point visual analogue scale with the 95% confidence interval (CI) at −1.04 to 2.84 —and no significant difference in investigator-rated eczema severity between probiotic and placebo treatments (Fig. 8.1) (Boyle et al. 2008). Subgroup analysis by age, eczema severity, presence of atopy, or presence of food allergy did not identify a population with different treatment outcomes (Boyle et al. 2008). Significant heterogeneity was noted between studies, and we found evidence that it may be explained by the use of different probiotic strains. Therefore, lack of effect based on pooled data from different probiotics does not exclude the possibility that a certain strain or strain combination

Table 8.1 Randomized controlled trials of probiotic ± prebiotic supplementation for the treatment of eczema

Study	Methods	Participants	Interventions	Outcomes
Brouwer 2006 (Brouwer et al. 2006)	3 month parallel group RCT	50 formula fed infants with eczema and suspected CMA	Test: eHF with *L. rhamnosus* or LGG at 5×10^9 cfu/100 ml Control: eHF without probiotic	SCORAD
Folster-Holst 2006 (Folster-Holst et al. 2006)	8 week parallel group RCT	53 children with eczema	Test: 1×10^{10} cfu/day LGG Control: microcrystalline cellulose	Global assessment; QoL score; SCORAD; Medication use
Gruber 2007 (Gruber et al. 2007)	12 week parallel group RCT	106 infants with eczema SCORAD 15–40	Test: 1×10^{10} cfu/day LGG Control: placebo capsule	SCORAD Medication use
Hattori 2003 (Hattori et al. 2003)	3 month parallel group RCT	17 children with eczema, suspected CMA, low fecal *Bifidobacterium* level	Test: eHF with prebiotic and *B. breve* M16-V at $5–15 \times 10^9$ cfu/day. Control: eHF with prebiotic and no probiotic	Eczema severity scoring scale
Isolauri 2000 (Isolauri et al. 2000)	Parallel group RCT ?duration	27 breast fed infants with eczema	Test: eHF with Bb-12 at 1×10^9 cfu/g or LGG at 3×10^8 cfu/g Control: eHF without probiotic	SCORAD
Kirjavainen 2003 (Folster-Holst et al. 2006)	Parallel group RCT ?duration	27 infants with eczema and suspected CMA	Test: eHF with LGG at 3×10^{10} cfu/kg/day Control: eHF without probiotic	SCORAD
Majamaa 1997 (Alfaleh and Bassler 2008)	1 month parallel group RCT	31 children with eczema and suspected CMA	Test: eHF with LGG at 5×10^8 cfu/g Control: eHF without probiotic	SCORAD
Passeron 2006 (Passeron et al. 2006)	3 month parallel group RCT	48 children with eczema SCORAD > 14	Test: prebiotic powder with *L. rhamnosus* Lcr35 at 3.6×10^9 cfu/day Control: prebiotic powder alone	Global assessment SCORAD

Rosenfeldt 2003 (Gueimonde et al. 2006)	6 week cross-over RCT	58 children with eczema	Test: *L. rhamnosus* 19070–2 and *L. reuteri* DSM12246 at 2×10^{10} cfu/day	Global assessment
			Control: skim milk powder + dextrose	Medication use
				SCORAD
Sistek 2006 (Sistek et al. 2006)	12 week parallel group RCT	60 children with eczema, atopy, SCORAD > 10	Test: *L. rhamnosus* and *B. lactis* at 2×10^{10} cfu/day	SCORAD
			Control: microcrystalline cellulose	
Viljanen 2005 (Viljanen et al. 2005a)	4 week parallel group RCT	252 infants with eczema and suspected CMA	Test: cow's milk elimination, eHF and LGG at 10^{10} cfu/day or probiotic mix	SCORAD
			Control: cow's milk elimination and eHF	
Weston 2005 (Weston et al. 2005)	8 week parallel group RCT	56 children with eczema	Test: *L. fermentum* VR1-003PCC 2×10^{9} cfu/day	Global assessment; QoL score; SCORAD; Medication use
		Modified SCORAD ≥ 25	Control: maltodextrin placebo	

Modified from Boyle et al. (2008)
RCT randomized controlled trial, *CMA* cow's milk allergy, *eHF* extensively hydrolyzed formula, *LGG Lactobacillus rhamnosus* GG, *CFU* colony-forming units, *SCORAD* Scoring Atopic Dermatitis, *QoL* quality of life

Study or Subgroup	Mean difference	SE	Weight	Mean difference IV, Random, 95% CI
1.3.1 Parallel group studies				
Folster-Holst 2006	3.7	4.102	13.3%	3.70 [-4.34, 11.74]
Gruber 2007	4.5	2.7296	16.5%	4.50 [-0.85, 9.85]
Passeron 2006	-3.28	5.3571	10.7%	-3.28 [-13.78, 7.22]
Sistek 2006	-15.47	5.0918	11.2%	-15.47 [-25.45, -5.49]
Viljanen 2005	2.78	1.8673	18.4%	2.78 [-0.88, 6.44]
Weston 2005	-8.18	4.3163	12.9%	-8.18 [-16.64, 0.28]
Subtotal (95% CI)			**83.0%**	**-1.68 [-7.15, 3.80]**
Heterogeneity: Tau² = 31.77; Chi² = 18.58, df = 5 (P = 0.002); I² = 73%				
Test for overall effect: Z = 0.60 (P = 0.55)				
1.3.2 Cross-over studies				
Rosenfeldt 2003	-6.2683	2.5226	17.0%	-6.27 [-11.21, -1.32]
Subtotal (95% CI)			**17.0%**	**-6.27 [-11.21, -1.32]**
Heterogeneity: Not applicable				
Test for overall effect: Z = 2.48 (P = 0.01)				
Total (95% CI)			**100.0%**	**-2.46 [-7.45, 2.53]**
Heterogeneity: Tau² = 31.86; Chi² = 24.77, df = 6 (P = 0.0004); I² = 76%				
Test for overall effect: Z = 0.97 (P = 0.33)				

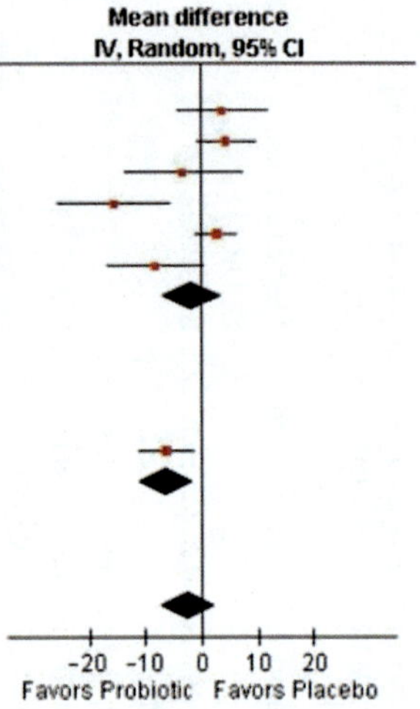

Fig. 8.1 Meta-analysis of published and unpublished data from randomized controlled trials (RCTs) of probiotics for the treatment of eczema. Of 12 published trials, 7 could be included in this meta-analysis, which shows the effect of probiotic treatment on eczema severity (SCORAD score, scale 0–102) at the end of treatment (Boyle et al. 2008)

could still be effective. In contrast to our meta-analysis, a recent meta-analysis by Michail et al. reported a significant difference favoring probiotics in reducing the SCORAD score of children with eczema (mean change from baseline was −3.01 with the 95% CI −5.36 to −0.66, P=0.01), and children with moderately severe disease were more likely to benefit (Michail et al. 2008). The authors of the latter meta-analysis were able to acquire original data from a greater number of included trials, which may account for the different conclusions of the two studies. However, the clinical significance of a three-point reduction in SCORAD score is unclear and compares poorly with the treatment effect seen with established eczema treatments. Topical application of probiotics for the treatment of eczema was recently evaluated in a single study that reported a significantly reduced SCORAD score and pruritus following treatment with topical *Vitreoscilla filiformis* lysate cream versus placebo and decreased loss of sleep compared to the start of treatment (Gueniche et al. 2008). The treatment was also associated with reduced *Staphylococcus aureus* colonization of skin. Future studies evaluating topical probiotics for the management of eczema will be of great interest.

8.3.2 Probiotics for Preventing Eczema

The trials evaluating early life oral probiotic administration for eczema prevention are more consistently positive than the eczema treatment studies. In all, 13 RCTs evaluating various probiotic bacteria used alone or in combination with other probiotics (and also prebiotics in one study) have now been reported (Abrahamsson et al. 2007; Dotterud et al. 2010; Huurre et al. 2008; Kalliomaki et al. 2001b; Kim et al. 2009; Kopp et al. 2008; Kukkonen et al. 2007; Niers et al. 2009; Rautava et al. 2006; Soh et al. 2009; Taylor et al. 2007; West et al. 2009; Wickens et al. 2008). Meta-analysis of these studies shows a significant protective effect of probiotics for preventing eczema and possibly IgE-associated eczema (Fig. 8.2a, b). Nine studies (evaluating 10 interventions) involved a combined prenatal (last 2–6 weeks of pregnancy) and postnatal (6–24 months) treatment: nine probiotic interventions (Abrahamsson et al. 2007; Dotterud et al. 2010; Huurre et al. 2008; Kalliomaki et al. 2001b; Kim et al. 2009; Kopp et al. 2008; Niers et al. 2009; Rautava et al. 2006; Soh et al. 2009; West et al. 2009; Wickens et al. 2008) and one synbiotic intervention (Kukkonen et al. 2007). All but two of these studies recruited infants at increased risk of allergic disease (first-degree relative with allergic disease): The studies by Huurre et al. (2008) and Dotterud et al. (2010) recruited an unselected population of mothers. Seven of the ten prenatal/postnatal treatments resulted in significantly less eczema during the first 2 years of life. A reduced cumulative incidence of eczema (Dotterud et al. 2010; Kalliomaki et al. 2001b; Kim et al. 2010), IgE-associated eczema (Abrahamsson et al. 2007), or both (Kukkonen et al. 2007; Wickens et al. 2008) at age 2 years was reported with six treatments; and one treatment resulted in a reduced cumulative incidence of parent-reported eczema and doctor-diagnosed eczema at age 3 months but not at 1 or 2 years of age, with no effect on IgE-associated eczema (Niers et al. 2009). Three of the ten prenatal/postnatal treatments showed no statistically significant reduction in rates of eczema or IgE-associated eczema at age 1 or 2 years. We performed a meta-analysis of studies that evaluated combined prenatal and postnatal treatment and found a significant protective effect with probiotic/prebiotic treatment for both eczema (Fig. 8.2c) and IgE-associated eczema (Fig. 8.2d) without a high level of heterogeneity between the results of individual studies.

In contrast to the findings for probiotics administered prenatally and postnatally, most of the studies (three of four) that evaluated postnatal without prenatal treatment with various probiotic bacteria reported no beneficial effects on the development of eczema or IgE-associated eczema at 12 months (Rautava et al. 2006; Soh et al. 2009; Taylor et al. 2007). Meta-analysis of these four studies revealed no evidence that postnatal treatment without a prenatal component reduces the risk of eczema (Fig. 8.2e). Moreover, in one study, postnatal treatment with *L. acidophilus* LAVRI-A1 was instead associated with an increased risk of IgE-associated eczema (RR 1.87) at 1 year (Taylor et al. 2007), although there was no significant difference between groups in atopic eczema prevalence at the 2.5-year follow-up (Prescott et al. 2008). One study reported a reduced cumulative incidence of eczema at 13 months following treatment with *L. paracasei* F19 during weaning (4–13 months of age) (West et al. 2009). Two of these postnatal studies recruited high-risk infants with a family history of allergic disease (Soh et al. 2009; Taylor et al. 2007), and the other two studies included formula-fed infants irrespective of a family history of allergic disease (Rautava et al. 2006; West et al. 2009).

Other allergic disease outcomes (recurrent wheeze, asthma, allergic rhinitis, food allergy) were assessed in five studies. Kukkonen et al. found no difference in the

a

Study or Subgroup	Probiotic Events	Probiotic Total	Placebo Events	Placebo Total	Weight	Risk Ratio M-H, Random, 95% CI
Abrahamsson 2007	34	95	32	93	9.1%	1.04 [0.70, 1.53]
Dotterud 2010	29	138	48	140	8.9%	0.61 [0.41, 0.91]
Huurre 2008	7	72	12	56	2.6%	0.45 [0.19, 1.08]
Kalliomaki 2001	15	64	31	68	6.2%	0.51 [0.31, 0.86]
Kim 2009	12	33	22	35	6.1%	0.58 [0.34, 0.97]
Kopp 2008	19	50	14	44	5.4%	1.19 [0.68, 2.09]
Kukkonen 2006	120	461	150	464	16.9%	0.81 [0.66, 0.99]
Niers 2009	23	50	30	48	9.6%	0.74 [0.51, 1.07]
Rautava 2006	4	32	8	40	1.7%	0.63 [0.21, 1.89]
Soh 2009	27	124	30	121	7.3%	0.88 [0.56, 1.39]
Taylor 2006	38	88	34	87	10.1%	1.10 [0.77, 1.58]
West 2009	9	84	19	87	3.5%	0.49 [0.24, 1.02]
Wickens 2008a	21	144	20	75	5.6%	0.55 [0.32, 0.94]
Wickens 2008b	37	152	20	75	7.1%	0.91 [0.57, 1.46]
Total (95% CI)		**1587**		**1433**	**100.0%**	**0.77 [0.66, 0.89]**
Total events	395		470			

Heterogeneity: Tau² = 0.02; Chi² = 19.00, df = 13 (P = 0.12); I² = 32%
Test for overall effect: Z = 3.51 (P = 0.0005)

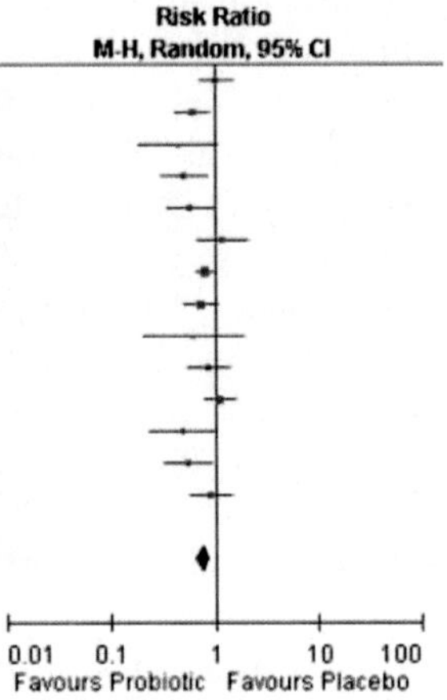

b

Study or Subgroup	Probiotic Events	Probiotic Total	Placebo Events	Placebo Total	Weight	Risk Ratio M-H, Random, 95% CI
Abrahamsson 2007	8	95	19	93	9.9%	0.41 [0.19, 0.89]
Dotterud 2010	9	129	10	130	8.5%	0.91 [0.38, 2.16]
Huurre 2008	7	72	12	56	8.6%	0.45 [0.19, 1.08]
Kim 2009	3	31	6	29	4.6%	0.47 [0.13, 1.70]
Kukkonen 2006	64	456	87	463	22.4%	0.75 [0.56, 1.00]
Niers 2009	10	50	8	48	8.9%	1.20 [0.52, 2.78]
Taylor 2006	23	88	12	86	12.7%	1.87 [1.00, 3.52]
Wickens 2008a	15	150	13	77	11.5%	0.59 [0.30, 1.18]
Wickens 2008b	20	152	14	77	12.8%	0.72 [0.39, 1.35]
Total (95% CI)		**1223**		**1059**	**100.0%**	**0.76 [0.56, 1.03]**
Total events	159		181			

Heterogeneity: Tau² = 0.08; Chi² = 13.96, df = 8 (P = 0.08); I² = 43%
Test for overall effect: Z = 1.75 (P = 0.08)

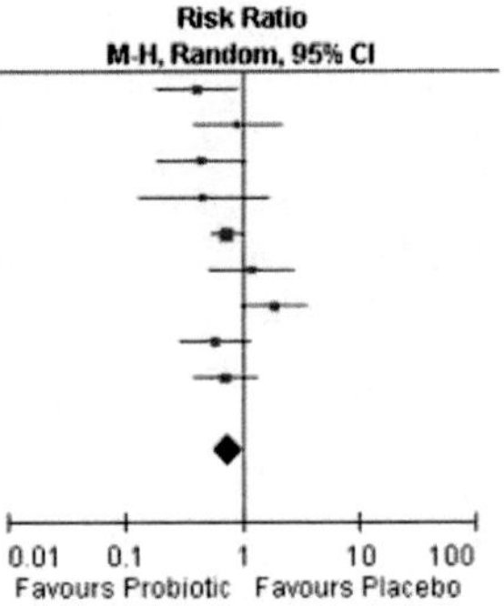

c

Study or Subgroup	Probiotic Events	Probiotic Total	Placebo Events	Placebo Total	Weight	Risk Ratio M-H, Random, 95% CI
Abrahamsson 2007	34	95	32	93	11.7%	1.04 [0.70, 1.53]
Dotterud 2010	29	138	48	140	11.4%	0.61 [0.41, 0.91]
Huurre 2008	7	72	12	56	3.3%	0.45 [0.19, 1.08]
Kalliomaki 2001	15	64	31	68	7.9%	0.51 [0.31, 0.86]
Kim 2009	12	33	22	35	7.8%	0.58 [0.34, 0.97]
Kopp 2008	19	50	14	44	6.9%	1.19 [0.68, 2.09]
Kukkonen 2006	120	461	150	464	22.4%	0.81 [0.66, 0.99]
Niers 2009	23	50	30	48	12.4%	0.74 [0.51, 1.07]
Wickens 2008a	21	144	20	75	7.2%	0.55 [0.32, 0.94]
Wickens 2008b	37	152	20	75	9.0%	0.91 [0.57, 1.46]
Total (95% CI)		**1259**		**1098**	**100.0%**	**0.74 [0.63, 0.87]**
Total events	317		379			

Heterogeneity: Tau² = 0.02; Chi² = 13.14, df = 9 (P = 0.16); I² = 32%
Test for overall effect: Z = 3.55 (P = 0.0004)

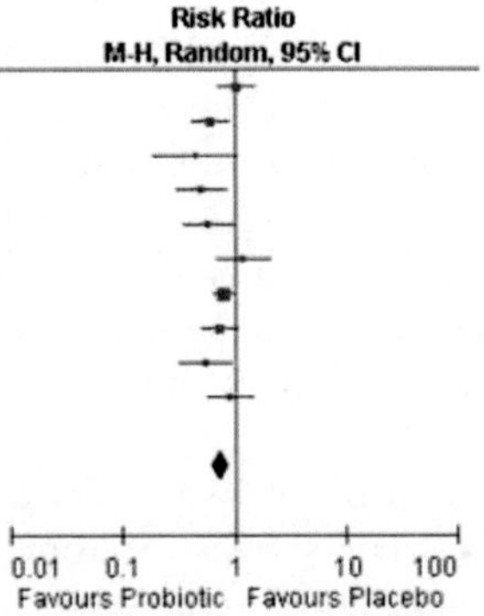

Fig. 8.2 (**a**) Meta-analysis of published data from RCTs of probiotics ± prebiotics for the prevention of eczema. (**b**) Meta-analysis of published data from RCTs of probiotics ± prebiotics for the prevention of immunoglobulin E (IgE)-associated eczema. (**c**) Meta-analysis of published data from RCTs of probiotics ± prebiotics for the prevention of eczema, where treatment was commenced prenatally. (**d**) Meta-analysis of published data from RCTs of probiotics ± prebiotics for the prevention of IgE-associated eczema, where treatment was commenced prenatally. (**e**) Meta-analysis of published data from RCTs of postnatal probiotics for the prevention of eczema, without a prenatal component to treatment

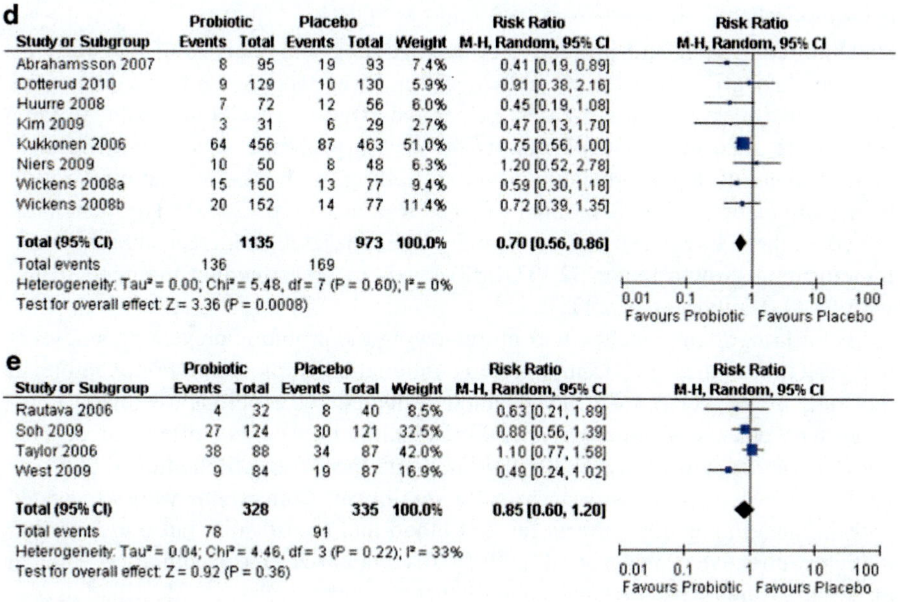

Fig. 8.2 (continued)

cumulative incidence of all allergic diseases or IgE-associated allergic diseases (Kukkonen et al. 2007), and Dotterud et al. found no difference in asthma or allergic rhinitis at age 2 years (Dotterud et al. 2010). However, Kopp et al. reported an almost threefold increased risk of recurrent wheezing bronchitis at age 2 years with LGG (Kopp et al. 2008) and trends to increased wheezing or asthma were noted in two other studies (Taylor et al. 2007; Kalliomaki et al. 2007). Additional data on asthma and allergic rhinitis outcomes will become available as further follow-up analyses are performed in the above studies.

Overall, these data suggest that probiotic treatment during late pregnancy and early postnatal life reduces the risk of eczema. It is necessary to examine the studies in a little more detail to define how best to administer probiotics in this context. First, we should note that most of the combined prenatal/postnatal treatments were effective in reducing eczema and/or IgE-associated eczema during the first 1–2 years of life, whereas most of the treatments involving only a postnatal component of therapy failed to reduce the risk for eczema and/or IgE-associated eczema. This suggests that a prenatal component of treatment is necessary for beneficial effects. Our own recently published study of prenatal probiotics without a postnatal component suggests that prenatal treatment is necessary, but not on its own sufficient for preventing eczema (Boyle et al. 2011).

Second, administration of a probiotic mix (LGG, *L. acidophilus* LA-5, *B. lactis* Bb-12) solely to women from 36 weeks of pregnancy to 3 months postpartum without direct infant supplementation was sufficient to reduce the cumulative incidence and prevalence of eczema at 2 years (Dotterud et al. 2010), indicating that direct infant probiotic supplementation during early life may not be an absolute requirement for protective effects. A subgroup analysis of the study by Kalliomaki's group also found evidence that direct supplementation of infants may not be necessary: the greatest protective effects of LGG treatment in their study were seen in breast-fed infants for whom probiotic was administered to their mothers during pregnancy and breastfeeding, without direct administration to the infant until after 3 months of age (Rautava et al. 2002). The beneficial effects in these breast-fed infants may have been related to increased breast milk transforming growth factor-β2 (TGFβ2) levels in breastfeeding mothers treated with LGG (Rautava et al. 2002).

A third important conclusion from reviewing the probiotic prevention studies is that selection of the right strain or strain combination is important. For example, in the study by Wickens et al. (2008), which evaluated two probiotic treatments compared with placebo, *L. rhamnosus* HN001 resulted in beneficial effects on eczema but *B. animalis* subsp *lactis* HN019 did not, and the difference in clinical outcome between the two strains was statistically significant. Both strains were associated with immunomodulatory effects in cord blood and breast milk, but only one was clinically effective (Prescott et al. 2008). This emphasizes the specificity of clinical effects of probiotic bacteria and suggests that the immune changes identified in cord blood and breast milk samples in that study may not be sufficient to prevent eczema. Thus, other mechanistic pathways need to be explored.

A fourth finding that is difficult to explain is the discrepant results of the studies by Kopp et al. and Kalliomaki et al. The two trials used a similar interventions of prenatal and postnatal LGG treatment for eczema prevention. The reasons for the discrepancy between the Kalliomaki and Kopp studies are not understood; however, genetic or dietary differences in the study populations may have contributed to these discrepant outcomes. It has been reported that gene polymorphisms in innate receptors may modify immune responses following signaling through these receptors. For example, CD14/-1721 polymorphisms can modulate the protective effects of farm milk ingestion on the development of eczema (Bieli et al. 2007). It is therefore possible that the discrepant outcomes of these two studies relate to genetic differences in the study populations. It is likely that the ability of microbial exposures (including probiotics and prebiotics) to modulate immune responses and protect against the development of allergic disease is influenced by individual genetic factors.

In summary, current studies suggest a potential role for probiotics or synbiotics in the prevention of eczema and/or IgE-associated eczema. Our meta-analyses (Fig. 8.2) suggest that probiotics/synbiotics are an effective intervention in this regard, particularly if treatment is administered both prenatally and postnatally. However, the very different nature of the interventions used in each trial published to date, combined with the inconsistent results of studies using the same intervention, make it difficult to translate these promising studies into a public health recommendation. Further work is needed to understand precisely which probiotic/synbiotic(s) are most effective for preventing eczema, in what way they should be given, and to gain a better understanding of the mechanism of action.

8.4 Probiotics for Other Skin Applications: Oral and Topical

A number of other potential skin applications for probiotics have been explored, although clinical trial data are much less complete than for studies in relation to eczema. The use of oral probiotics for promoting immune responses in the skin has been explored in relation to UV light-induced skin damage, where one clinical study suggested that probiotic treatment may have a role in promoting recovery from such damage in those most susceptible to it (Peguet-Navarro et al. 2008). In vitro work has identified specific effects of probiotics on skin cells in culture, which may be associated with cosmetic benefits, although clinical studies are clearly needed to confirm this possibility (Baba et al. 2006; Miyazaki et al. 2003). Topical use of probiotics for the treatment of eczema, cosmetic indications, and wound healing have also been explored. Although there is little evidence that this mode of treatment would lead to any long-term modulation of the skin microbiota, it may nevertheless lead to long-term immune or skin barrier changes. This has been formally examined in the area of wound healing, with promising preliminary results (Nikitenko 2004; Rodrigues et al. 2005; Valdez et al. 2005). The optimal approach for topical probiotic treatment may be the use of probiotic metabolites (sometimes called postbiotics) because, in contrast to oral administration where such metabolites may be degraded in the upper gastrointestinal tract, direct skin application of metabolites avoids these issues and may be safer than applying a live probiotic bacterium to an open wound. There is clearly considerable room for further investigation of topical probiotic and postbiotic treatments for skin applications.

8.5 Conclusion

Studies suggest a potential role for selected probiotics in the prevention of eczema (Osborn and Sinn 2007). Our own meta-analyses (Fig. 8.2) suggest that a prenatal component of treatment is important for beneficial effects. It may be important to continue treatment during the postnatal period for maximum effect, and indirect administration to mothers while they are breastfeeding may be especially beneficial. In contrast, administration of probiotics solely during the postnatal period has not proved beneficial (Fig. 8.2e). To translate the promising findings regarding probiotic intervention for eczema prevention into a meaningful public health intervention, further work is needed to clarify the optimal dose and combinations of bacterial strains, whether there is added benefit in combining probiotic with prebiotic, the optimal timing for intervention, and the patient populations who would most benefit from such therapy. Studies examining the use of probiotics for the treatment of allergic disease have been disappointing. A recent Cochrane systematic review (Boyle et al. 2008) concluded that probiotics were not effective for the treatment of eczema. Studies of probiotics for other skin conditions are at a much earlier stage of development but show promise, particularly in the augmentation of wound healing. The application of topical probiotics to treat eczema and other skin conditions is a recent development that warrants further study in the future.

Take Home Messages

- The term probiotic encompasses any live microorganism with a proven health benefit.
- Clinical trials, rather than animal or in vitro work, are critical for identifying which microbes have probiotic properties.
- The intense microbial colonization of human epithelial surfaces may be influenced by the use of probiotics, so they may have benefits for patients with skin and intestinal diseases.
- Current evidence regarding probiotic effects on skin disease is most positive for eczema prevention.
- Topical use of probiotics is a promising approach, but further studies are required to clarify the effects of topical probiotics on skin health.

References

Abrahamsson TR, Jakobsson T, Bottcher MF, Fredrikson M, Jenmalm MC, Bjorksten B, et al. Probiotics in prevention of IgE-associated eczema: a double-blind, randomized, placebo-controlled trial. J Allergy Clin Immunol. 2007;119(5):1174–80.

Alander M, Satokari R, Korpela R, Saxelin M, Vilpponen-Salmela T, Mattila-Sandholm T, et al. Persistence of colonization of human colonic mucosa by a probiotic strain, *Lactobacillus rhamnosus* GG, after oral consumption. Appl Environ Microbiol. 1999;65(1):351–4.

Alfaleh K, Bassler D. Probiotics for prevention of necrotizing enterocolitis in preterm infants. Cochrane Database Syst Rev. 2008;1:CD005496.

Allen SJ, Okoko B, Martinez E, Gregorio G, Dans LF. Probiotics for treating infectious diarrhoea. Cochrane Database Syst Rev. 2003;4:CD003048.

Aragon G, Graham DB, Borum M, Doman DB. Probiotic therapy for irritable bowel syndrome. Gastroenterol Hepatol. 2010;6(1):39–44.

Baba H, Masuyama A, Takano T. Short communication: effects of *Lactobacillus helveticus*-fermented milk on the differentiation of cultured normal human epidermal keratinocytes. J Dairy Sci. 2006;89(6):2072–5.

Bek-Thomsen M, Lomholt HB, Kilian M. Acne is not associated with yet-uncultured bacteria. J Clin Microbiol. 2008;46(10):3355–60.

Bieli C, Eder W, Frei R, Braun Fahrlander C, Klimecki W, Waser M, et al. A polymorphism in CD14 modifies the effect of farm milk consumption on allergic diseases and CD14 gene expression. J Allergy Clin Immunol. 2007;120:1308–15.

Bjorksten B, Sepp E, Julge K, Voor T, Mikelsaar M. Allergy development and the intestinal microflora during the first year of life. J Allergy Clin Immunol. 2001;108(4):516–20.

Boyle RJ, Bath-Hextall FJ, Leonardi-Bee J, Murrell DF, Tang ML. Probiotics for treating eczema. Cochrane Database Syst Rev. 2008;4:CD006135.

Boyle RJ, Ismail IH, Kivivuori S, Licciardi PV, Robins-Browne RM, Mah LJ, et al. Lactobacillus GG treatment during pregnancy for the prevention of eczema: a randomized controlled trial. Allergy 2011;66(4):509–16.

Brouwer ML, Wolt-Plompen SA, Dubois AE, van der Heide S, Jansen DF, Hoijer MA, et al. No effects of probiotics on atopic dermatitis in infancy: a randomized placebo-controlled trial. Clin Exp Allergy. 2006;36(7):899–906.

D'Souza AL, Rajkumar C, Cooke J, Bulpitt CJ. Probiotics in prevention of antibiotic associated diarrhoea: meta-analysis. BMJ. 2002;324(7350):1361.

Dekio I, Sakamoto M, Hayashi H, Amagai M, Suematsu M, Benno Y. Characterization of skin microbiota in patients with atopic dermatitis and in normal subjects using 16 S rRNA gene-based comprehensive analysis. J Med Microbiol. 2007;56(Pt 12):1675–83.

Dotterud CK, Storro O, Johnsen R, Oien T. Probiotics in pregnant women to prevent allergic disease: a randomised, double-blind trial. Br J Dermatol. 2010;163(3):616–23.

FAO/WHO. Guidelines for the evaluation of probiotics in food. London, Ontario, Canada; 2002.

Folster-Holst R, Muller F, Schnopp N, Abeck D, Kreiselmaier I, Lenz T, et al. Prospective, randomized controlled trial on *Lactobacillus rhamnosus* in infants with moderate to severe atopic dermatitis. Br J Dermatol. 2006;155(6):1256–61.

Gao Z, Tseng CH, Strober BE, Pei Z, Blaser MJ. Substantial alterations of the cutaneous bacterial biota in psoriatic lesions. PLoS One. 2008;3(7):e2719.

Gionchetti P, Rizzello F, Venturi A, Brigidi P, Matteuzzi D, Bazzocchi G, et al. Oral bacteriotherapy as maintenance treatment in patients with chronic pouchitis: a double-blind, placebo-controlled trial. Gastroenterology. 2000;119(2):305–9.

Gruber C, Wendt M, Sulser C, Lau S, Kulig M, Wahn U, et al. Randomized, placebo-controlled trial of *Lactobacillus rhamnosus* GG as treatment of atopic dermatitis in infancy. Allergy. 2007;62(11):1270–6.

Gueimonde M, Sakata S, Kalliomaki M, Isolauri E, Benno Y, Salminen S. Effect of maternal consumption of lactobacillus GG on transfer and establishment of fecal bifidobacterial microbiota in neonates. J Pediatr Gastroenterol Nutr. 2006;42(2):166–70.

Gueniche A, Knaudt B, Schuck E, Volz T, Bastien P, Martin R, et al. Effects of nonpathogenic gram-negative bacterium *Vitreoscilla filiformis* lysate on atopic dermatitis: a prospective, randomized, double-blind, placebo-controlled clinical study. Br J Dermatol. 2008;159(6):1357–63.

Guillemard E, Tondu F, Lacoin F, Schrezenmeir J. Consumption of a fermented dairy product containing the probiotic *Lactobacillus casei* DN-114001 reduces the duration of respiratory infections in the elderly in a randomised controlled trial. Br J Nutr. 2010;103(1): 58–68.

Hattori K, Yamamoto A, Sasai M, Taniuchi S, Kojima T, Kobayashi Y, et al. Effects of administration of bifidobacteria on fecal microflora and clinical symptoms in infants with atopic dermatitis. Arerugi. 2003;52(1):20–30.

Huurre A, Laitinen K, Rautava S, Korkeamaki M, Isolauri E. Impact of maternal atopy and probiotic supplementation during pregnancy on infant sensitization: a double-blind placebo-controlled study. Clin Exp Allergy. 2008;38(8):1342–8.

Isolauri E, Arvola T, Sutas Y, Moilanen E, Salminen S. Probiotics in the management of atopic eczema. Clin Exp Allergy. 2000;30(11):1604–10.

Kaila M, Isolauri E, Saxelin M, Arvilommi H, Vesikari T. Viable versus inactivated lactobacillus strain GG in acute rotavirus diarrhoea. Arch Dis Child. 1995;72(1):51–3.

Kalliomaki M, Kirjavainen P, Eerola E, Kero P, Salminen S, Isolauri E. Distinct patterns of neonatal gut microflora in infants in whom atopy was and was not developing. J Allergy Clin Immunol. 2001a;107(1):129–34.

Kalliomaki M, Salminen S, Arvilommi H, Kero P, Koskinen P, Isolauri E. Probiotics in primary prevention of atopic disease: a randomised placebo-controlled trial. Lancet. 2001b;357(9262): 1076–9.

Kalliomaki M, Salminen S, Poussa T, Isolauri E. Probiotics during the first 7 years of life: a cumulative risk reduction of eczema in a randomized, placebo-controlled trial. J Allergy Clin Immunol. 2007;119(4):1019–21.

Kim JY, Kwon JH, Ahn SH, Lee SI, Han YS, Choi YO, et al. Effect of probiotic mix (*Bifidobacterium bifidum, Bifidobacterium lactis, Lactobacillus acidophilus*) in the primary prevention of eczema: a double-blind, randomized, placebo-controlled trial. Pediatr Allergy Immunol. 2009;21(2 Pt 2):e386–93.

Kim JY, Kwon JH, Ahn SH, Lee SI, Han YS, Choi YO, et al. Effect of probiotic mix (*Bifidobacterium bifidum, Bifidobacterium lactis, Lactobacillus acidophilus*) in the primary prevention of eczema: a double-blind, randomized, placebo-controlled trial. Pediatr Allergy Immunol. 2010;21(2 Pt 2):e386–93.

Kirjavainen PV, Salminen SJ, Isolauri E. Probiotic bacteria in the management of atopic disease: underscoring the importance of viability. J Pediatr Gastroenterol Nutr. 2003;36(2):223–7.

Kopp MV, Hennemuth I, Heinzmann A, Urbanek R. Randomized, double-blind, placebo-controlled trial of probiotics for primary prevention: no clinical effects of Lactobacillus GG supplementation. Pediatrics. 2008;121(4):e850–6.

Kukkonen K, Savilahti E, Haahtela T, Juntunen-Backman K, Korpela R, Poussa T, et al. Probiotics and prebiotic galacto-oligosaccharides in the prevention of allergic diseases: a randomized, double-blind, placebo-controlled trial. J Allergy Clin Immunol. 2007;119(1):192–8.

Lahtinen SJ, Boyle RJ, Kivivuori S, Oppedisano F, Smith KR, Robins-Browne R, et al. Prenatal probiotic administration can influence *Bifidobacterium microbiota* development in infants at high risk of allergy. J Allergy Clin Immunol. 2009;123(2):499–501.

Leyer GJ, Li S, Mubasher ME, Reifer C, Ouwehand AC. Probiotic effects on cold and influenza-like symptom incidence and duration in children. Pediatrics. 2009;124(2):e172–9.

Majamaa H, Isolauri E. Probiotics: a novel approach in the management of food allergy. J Allergy Clin Immunol. 1997;99(2):179–85.

Metchnikoff E. Prolongation of life: optimistic studies. New York: Putnam and Sons; 1906.

Michail SK, Stolfi A, Johnson T, Onady GM. Efficacy of probiotics in the treatment of pediatric atopic dermatitis: a meta-analysis of randomized controlled trials. Ann Allergy Asthma Immunol. 2008;101(5):508–16.

Miyazaki K, Hanamizu T, Iizuka R, Chiba K. Bifidobacterium-fermented soy milk extract stimulates hyaluronic acid production in human skin cells and hairless mouse skin. Skin Pharmacol Appl Skin Physiol. 2003;16(2):108–16.

Niers L, Martin R, Rijkers G, Sengers F, Timmerman H, van Uden N, et al. The effects of selected probiotic strains on the development of eczema (the PandA study). Allergy. 2009;64(9):1349–58.

Nikitenko VI. Infection prophylaxis of gunshot wounds using probiotics. J Wound Care. 2004;13(9):363–6.

Osborn DA, Sinn JK. Probiotics in infants for prevention of allergic disease and food hypersensitivity. Cochrane Database Syst Rev. 2007;4:CD006475.

Passeron T, Lacour JP, Fontas E, Ortonne JP. Prebiotics and synbiotics: two promising approaches for the treatment of atopic dermatitis in children above 2 years. Allergy. 2006;61(4):431–7.

Peguet-Navarro J, Dezutter-Dambuyant C, Buetler T, Leclaire J, Smola H, Blum S, et al. Supplementation with oral probiotic bacteria protects human cutaneous immune homeostasis after UV exposure-double blind, randomized, placebo controlled clinical trial. Eur J Dermatol. 2008;18(5):504–11.

Penders J, Thijs C, van den Brandt PA, Kummeling I, Snijders B, Stelma F, et al. Gut microbiota composition and development of atopic manifestations in infancy: the KOALA Birth Cohort study. Gut. 2007;56(5):661–7.

Prescott SL, Wickens K, Westcott L, Jung W, Currie H, Black PN, et al. Supplementation with *Lactobacillus rhamnosus* or *Bifidobacterium lactis* probiotics in pregnancy increases cord blood interferon-gamma and breast milk transforming growth factor-beta and immunoglobin A detection. Clin Exp Allergy. 2008;38(10):1606–14.

Rautava S, Kalliomaki M, Isolauri E. Probiotics during pregnancy and breast-feeding might confer immunomodulatory protection against atopic disease in the infant. J Allergy Clin Immunol. 2002;109(1):119–21.

Rautava S, Arvilommi H, Isolauri E. Specific probiotics in enhancing maturation of IgA responses in formula-fed infants. Pediatr Res. 2006;60(2):221–4.

Rodrigues KL, Caputo LR, Carvalho JC, Evangelista J, Schneedorf JM. Antimicrobial and healing activity of kefir and kefiran extract. Int J Antimicrob Agents. 2005;25(5):404–8.

Roessler A, Friedrich U, Vogelsang H, Bauer A, Kaatz M, Hipler UC, et al. The immune system in healthy adults and patients with atopic dermatitis seems to be affected differently by a probiotic intervention. Clin Exp Allergy. 2008;38(1):93–102.

Rosenfeldt V, Benfeldt E, Nielsen SD, Michaelsen KF, Jeppesen DL, Valerius NH, et al. Effect of probiotic Lactobacillus strains in children with atopic dermatitis. J Allergy Clin Immunol. 2003;111(2):389–95.

Sistek D, Kelly R, Wickens K, Stanley T, Fitzharris P, Crane J. Is the effect of probiotics on atopic dermatitis confined to food sensitized children? Clin Exp Allergy. 2006;36(5):629–33.

Soh SE, Aw M, Gerez I, Chong YS, Rauff M, Ng YP, et al. Probiotic supplementation in the first 6 months of life in at risk Asian infants–effects on eczema and atopic sensitization at the age of 1 year. Clin Exp Allergy. 2009;39(4):571–8.

Sudo N, Sawamura S, Tanaka K, Aiba Y, Kubo C, Koga Y. The requirement of intestinal bacterial flora for the development of an IgE production system fully susceptible to oral tolerance induction. J Immunol. 1997;159(4):1739–45.

Taylor AL, Dunstan JA, Prescott SL. Probiotic supplementation for the first 6 months of life fails to reduce the risk of atopic dermatitis and increases the risk of allergen sensitization in high-risk children: a randomized controlled trial. J Allergy Clin Immunol. 2007;119(1):184–91.

Valdez JC, Peral MC, Rachid M, Santana M, Perdigon G. Interference of *Lactobacillus plantarum* with *Pseudomonas aeruginosa* in vitro and in infected burns: the potential use of probiotics in wound treatment. Clin Microbiol Infect. 2005;11(6):472–9.

Viljanen M, Savilahti E, Haahtela T, Juntunen-Backman K, Korpela R, Poussa T, et al. Probiotics in the treatment of atopic eczema/dermatitis syndrome in infants: a double-blind placebo-controlled trial. Allergy. 2005a;60(4):494–500.

Viljanen M, Pohjavuori E, Haahtela T, Korpela R, Kuitunen M, Sarnesto A, et al. Induction of inflammation as a possible mechanism of probiotic effect in atopic eczema-dermatitis syndrome. J Allergy Clin Immunol. 2005b;115(6):1254–9.

Watanabe S, Narisawa Y, Arase S, Okamatsu H, Ikenaga T, Tajiri Y, et al. Differences in fecal microflora between patients with atopic dermatitis and healthy control subjects. J Allergy Clin Immunol. 2003;111(3):587–91.

West CE, Hammarstrom ML, Hernell O. Probiotics during weaning reduce the incidence of eczema. Pediatr Allergy Immunol. 2009;20(5):430–7.

Weston S, Halbert AR, Richmond P, Prescott SL. Effects of probiotics on atopic dermatitis: a randomised controlled trial. Arch Dis Child. 2005;90(9):892–7.

Wickens K, Black PN, Stanley TV, Mitchell E, Fitzharris P, Tannock GW, et al. A differential effect of 2 probiotics in the prevention of eczema and atopy: a double-blind, randomized, placebo-controlled trial. J Allergy Clin Immunol. 2008;122(4):788–94.

Part II
Clinical Crossover Between Nutrition and Dermatology

Chapter 9
Diet and Acne

Apostolos Pappas

Core Messages

- Nutrition may play a role in acne development, especially a diet with high glycemic load.
- Given that acne has a complex etiology of several factors, it seems unlikely that one nutrient could be responsible for its clinical manifestations.
- Retinoids and carotenoids, which are, respectively, metabolites and precursors of vitamin A, target skin cells.
- The role of dairy foods in acne development has not been fully validated, and proper clinical studies are needed to address this role.
- Few clinical studies have successfully addressed the relationship between diet and acne.

9.1 Introduction

Nutrition and diet affect overall health and general well-being. This statement needs no particular citation because every nutritional textbook advocates it. But could diet affect acne? Acne is one of the most common dermatological conditions, affecting millions of people worldwide (Thiboutot 2008). It is generally accepted that excess sebaceous lipids, hormones, bacteria, and hyperproliferation of follicular cells are the major etiological factors for acne (Zouboulis 2004).

A. Pappas (✉)
The Johnson & Johnson Skin Research Center, CPPW, A Division of Johnson & Johnson Consumer Companies, Inc., Grandview Road 199, Skillman, NJ 08558, USA
e-mail: apostolos_pappas@yahoo.com

A. Pappas (ed.), *Nutrition and Skin: Lessons for Anti-Aging, Beauty and Healthy Skin*,
DOI 10.1007/978-1-4419-7967-4_9,

The current status of the relation between diet and acne is unclear and under debate. On one hand, the American Academy of Dermatology (AAD) published recommendations (Strauss et al. 2007) in 2007 suggesting that caloric restriction has no benefit in the treatment of acne and that there is insufficient evidence to link the consumption of certain "food enemies" to acne (http www skincarephysicians com acnenet acne_and_diet html). On the other hand, recent clinical studies have suggested a rather close relation between diet and acne (Smith et al. 2007a, 2008).

Let us start from the beginning and, in particular, from the founder of modern medicine, Hippocrates. One of his main principles and teachings was "Let food be your medicine, and let medicine be your food." This statement was quoted in another review on the subject of diet and acne (Wolf et al. 2004). Since that review was published in 2004, many articles and commentaries have been published on that relationship (Danby 2008, 2009; Cordain 2005; Treloar et al. 2008; Webster 2008; Treloar 2008). Before continuing with the review of the publications of the last few years, it is appropriate to quote the conclusions of that review because, in my opinion, there is no better way to express the present state of knowledge.

> We did not realize how daunting it would be to write an article dedicated to making sense of the relationship of acne to foods. It turns out that there are no meta-analyses, randomized controlled clinical studies, or well designed scientific trials that follow evidence-based guidelines for providing solid proof in dealing with this issue.
>
> We emerged from our search disappointed and confess at the outset that what we present in this article will not settle this controversial issue and that the reader will not get a clear-cut message from us; such is the nature of the beast.
>
> We reviewed the updated arguments, facts, and relevant data on this ancient debate, but we warn the truth-seekers among you that the jury is still out.

Indeed, at that time there were no better words to describe "the nature of the beast."

9.2 Could Certain Essential Nutrients Affect Acne?

It is obvious that Hippocrates' ancient but wise statement should hold some truth when applied to acne, given that the most efficacious current therapies for acne are retinoids. Oral administration of isotretinoin (13-*cis*-retinoic acid, or Accutane) and topical application of its isomers and natural retinoids (e.g. tretinoin) are used as anti-acne therapies (Webster et al. 2009; Berger et al. 2007). 13-*cis*-retinoic acid (RA) is the only drug that targets all four pathogenic factors of acne and is the most efficient so far in regard to sebum suppression (Katsambas and Dessinioti 2008). 13-*cis*-RA is a retinoid that potentially derives from the metabolism of vitamin A. Although several websites state that it is found in small quantities naturally in the body without citing a reference (Vahlquist 1999), we know that at least the natural isomers of RA also affect the disease. With that in mind, we can undoubtedly predict an association between diet and acne.

Vitamin A is essential for skin's health. Vitamin A deficiency causes abnormal visual adaptation to darkness (night blindness) but also dramatically affects cutaneous biology.

Dry skin, dry hair, and broken fingernails are among the first manifestations of vitamin A deficiency (Russell 2001). This nutrient, which is stored in the liver, is found also in the skin, particularly in the sebaceous glands, which express retinoid receptors (Orfanos et al. 1997; Tsukada et al. 2000). Let us also not forget that most dermatologists nowadays recommend ingestion of isotretinoin with fatty foods. This recommendation stems from nutritional studies on carotenoids. Retinol (vitamin A), carotenoids (provitamin A), and subsequently retinoids (vitamin A metabolites) are absorbed better with parallel intake of vegetable oils (Fielding et al. 2005; Unlu et al. 2005; Brown et al. 2004; Harrison 2005; Mulokozi et al. 2004; Ribaya-Mercado 2002).

Therefore, how could the possibility that diet has no effect on acne could be ruled out, especially when diet influences the absorbance of a nutrient, its metabolites, and a drug that affects mitigation of that disease? Maybe acne cannot be cured with nutrition, but diet could certainly influence the status of the disease. Perhaps food does not cause or eradicate the disease, but certainly it could ameliorate or worsen its severity.

Let us also consider that lipophilic vitamins A and D have an important impact on keratinocyte biology, which can be crucial to their proliferation in acne. The susceptibility of keratinocytes to the antiproliferative effects of vitamins A and D has been documented (Popadic et al. 2008). In vitro studies have shown that RA inhibits proliferation of mouse and human keratinocytes via a peroxisome proliferator-activated receptor (PPAR) independent mechanism (Borland et al. 2008). Another report demonstrated that all-*trans* RA, an isoform of RA, increased aquaporin-3 expression and enhanced its biological activity in human skin (Bellemère et al. 2008).

Vitamins A and D are the initial group of nutrients that have been reported to exhibit properties of skin hormones (Reichrath et al. 2007). They control metabolism, inactivation, activation, and elimination of specialized skin cells. Many retinoids are also hormones as they bind to and activate specific nuclear receptors, affecting their function. Vitamin A and its natural metabolites have been approved for topical and systemic treatment of mild, moderate, and severe, recalcitrant acne, as well as photoaging, biological skin aging, acute promyelocytic leukemia, and Kaposi's sarcoma (Reichrath et al. 2007). Vitamin D's critical importance for the skin and consequently the body's endocrine system is demonstrated by the fact that the skin is the site where active vitamin D metabolites, such as $1,25(OH)_2D_3$, are synthesized. In keratinocytes, $1,25(OH)_2D_3$ regulates growth and differentiation; and for this reason vitamin D analogues have been developed for the treatment of psoriasis, an aggressive hyperproliferative skin disease. In addition vitamin D analogues are reported to affect the immune system and to offer protection against cancer and autoimmune and infectious diseases in various organs and tissues (Reichrath et al. 2007). All the above examples are cited to reiterate the fact that these nutrients and their metabolites could influence skin hydration, hyperproliferation, and metabolism.

In addition to vitamins A and D, reports have proven that the other major lipophilic vitamin, E, is delivered onto the skin via sebaceous gland activity (Thiele et al. 1999; Ekanayake-Mudiyanselage et al. 2004). Such sebaceous delivery could make a difference in inflammatory acne because lipid oxidation could increase the inflammation status of the disease. Antinflammatory compounds such as zileuton, which targets certain enzymes of the lipid oxidation pathways, have been shown to reduce inflammatory lesion counts and sebum production (Zouboulis et al. 2003, 2005). These pathways involve metabolites of polyunsaturated fatty acids (PUFAs). There are also two fatty acids in our body that are essential and cannot be synthesized by human cells: linoleic acid (18:2, Δ9,12) and linolenic acid (18:2, Δ9,12,15) (LA). These are important nutrients that need to be obtained from the diet, which is why they are referred to as *essential* fatty acids. These two essential nutrients are precursors to the omega-6 and omega-3 fatty acid families, respectively, a family of metabolites that are involved in numerous important physiological processes, including inflammation. Consequently we could safely assume that the absence of these important nutrients from our diet could have important implications for both acne and our overall health. Numerous studies have revealed that clinical imbalances of specific essential fatty acids are associated with a variety of skin problems, such as dry, itchy, scaly skin, which is a hallmark sign of fatty acid deficiency (Horrobin 1989). More relevant to this is a publication that suggested that the sebum of acne patients is relatively deficient in linoleic acid (Downing et al. 1986).

The exact fate of these essential nutrients in human sebaceous cells is not yet entirely elucidated. An experimental study (Pappas et al. 2002) unveiled a unique metabolic fate of linoleic acid in sebaceous cells, which is preferentially β-oxidized, contrary to the other predominant fatty acids, which are incorporated to the most common sebaceous lipids. That rapid oxidation and degradation in sebaceous cells allows palmitic acid to be available as the sole substrate to the δ6 desaturase of sebaceous cells, which is the predominant desaturase of human sebaceous cells (Ge et al. 2003). That enzyme usually catalyzes the synthesis of more omega-6 derivatives from linoleic acid, since it is the enzyme's preferred substrate. There is also considerable evidence that linoleic acid is an essential structural component of skin ceramides, important for barrier function.

Sebum analysis demonstrates that these essential fatty acids and their derivatives comprise small amounts of surface lipids (Nicolaides 1974). However, two intriguing studies (Fu and Sinclair 2000; Fu et al. 2001) revealed a firm association of these two fatty acids and the skin. When guinea pigs consumed radioactively labeled linoleic and α-LA, the lipids in skin and fur were predominantly labeled.

Following the administration of ^{14}C-labeled αLA, 46% of the radioactivity was associated with the skin and fur lipids; and about 39% of the label was not recovered in the body lipids and was assumed to have been expired as CO_2 or unabsorbed. These data identify a new route of metabolism of αLA in these species and presumably through the sebaceous glands onto fur lipids and subsequently skin. Perhaps in humans the distribution could be different, but at least the above study revealed that these essential nutrients could enter from the diet, survive the digestive tract, and reach the skin's surface unaltered. A recent nutritional clinical study (De Spirt et al. 2009) in two groups of women who consumed flaxseed or borage oil for 12 weeks revealed that the daily ingestion of 2.2 g αLA or 2.2 g linoleic and γ-linolenic acid, respectively, demonstrated skin benefits. Skin irritation, changes in skin reddening, and blood flow were diminished in both groups compared to the placebo group, providing evidence that skin properties can be modulated by intervention with dietary fatty acids.

Another class of nutrients that derives from the diet includes minerals such as zinc, copper, selenium, and iron, which known to influence antiinflammatory and proinflammatory enzymes (e.g., desaturases and lipoxygenases). Could a diet rich in zinc or selenium bring benefit to a patient with acne? Could a diet rich in iron or copper worsen acne? We simply do not know because the proper nutritional clinical studies have not been performed.

9.3 Could Diet Affect Acne Status?

The fact that Western diets are often deficient in the longer-chain omega-3 s and their precursor αLA raises an additional issue for this discussion. It is known that the ratio of omega-6 to omega-3 fatty acids in a typical western diet ranges from about 10:1 to 20:1 (Kris-Etherton et al. 2000; Logan 2003) versus a ratio of 3:1–1:1 in a non-Western diet (Logan 2003) or in primitive, nonindustrialized populations (Cordain et al. 2002).

These findings were the foundation for population studies that revealed that non-Western diets correlated with the absence of acne (Cordain et al. 2002). A number of studies have suggested that inflammatory markers correlate with an increase of the omega-6/omega-3 ratio (Kris-Etherton et al. 2000). The omega-6 fatty acids are thought to induce more proinflammatory mediators and have been associated with the development of inflammatory acne (Zouboulis 2001; Trebble et al. 2003). On the other hand, intake of high levels of omega-3 fatty acids is linked with decreases in inflammatory factors (James et al. 2000). In addition, there are epidemiological studies demonstrating that increasing the intake of omega-3 fatty acids through a diet rich in seafood results in lower rates of inflammatory disease (Kris-Etherton et al. 2000; Rubin et al. 2008). There are also studies claiming that sebum production is increased by the consumption of dietary fat or carbohydrates (Llewellyn 1967) and that variations in carbohydrates could also influence sebum composition (MacDonald 1964, 1967).

In general, our Western diet is not just deprived of omega-3 s but is also a diet rich in refined carbohydrates. It has been reported that people living in the Kitavan Islands (off the coast of Papua New Guinea) and the Aché hunter-gathers of Paraguay do not suffer from acne and that it is associated with their low-glycemic diet, consisting mainly of fresh vegetables, fruits, and lean proteins (Cordain et al. 2002). This conclusion is in agreement with the latest studies (Smith et al. 2007a, 2008) on low-glycemic diets (discussed later). In brief, one prospective cohort study (Smith et al. 2007b) found a valid association between high-glycemic-index foods and longer acne duration, whereas two randomized controlled trials (Smith et al. 2007a, 2008) associated low-glycemic-index diet with reduced acne risk.

In addition to these reports, two previous studies had reported on how caloric restriction can alter sebum composition. However, we do not know if this could relate directly to the pathological condition of acne (Downing et al. 1972; Pochi et al. 1970).

9.4 Status and Studies

As noted above, a comprehensive review of the literature in 2004 concluded that there was no definitive evidence on the effects of diet on acne. Has there been any progress since then? A year later, another review (Cordain 2005) restated that within the dermatological community a consensus had emerged that diet was unrelated to the etiology of acne. That review summarized the few poorly designed studies, more than three decades old, which contained few objective data or analyses. In general, those studies were inconclusive due to various methodological limitations: small sample size, lack of appropriate controls, potential recall bias, incomplete reported results, or failure to define clearly the changes in acne (Anderson 1971; Fulton et al. 1969).

Interestingly, the same review mentioned that there should be a link between diet and acne as many dietary factors influence a variety of hormones and growth factors that, in turn, influence sebaceous gland biology and synthesis of sebum. At the end of the article, the authors noted that there had been no recent studies to explore the relation of diet and acne.

That same year, a study (Adebamowo et al. 2005) was published that linked acne to the consumption of milk and especially skim milk. The authors raised also the point that most of the milk and dairy products consumed in the United States come from pregnant cows. Could they be responsible for acne as milk exposes us to the hormones that cows produce during pregnancy? Given also the fact that hormones play an obvious role in acne, sebum production may be influenced by androgens and hormonal mediators, including sex hormone-binding globulin (SHBG) and insulin-like growth factor-1 (IGF-I), all of which may be influenced by dietary factors. The study was based on a questionnaire sent to a group of 47,355 women who were asked to remember what they ate in high school, years prior to the study. Another later study from the same team asked teenage boys to recall what they ate and to self-determine the severity of their acne (Adebamowo et al. 2008). The investigators concluded that there was an association between drinking milk and acne. However these studies had major limitations because the questionnaire required self-assessment of acne and was based on memory of food intake. This could be difficult and subjective because recalling what one ate days ago can be difficult. In addition, an association between drinking milk and acne means that more validated and well-designed studies are needed to determine if there is an association or a cause.

Factors such as heredity were ignored, and the data revealed a low prevalence rate of acne across every group. For example, the group that consumed two to three glasses of milk per day had 1344 responders that reported acne—only 7.7% of the total of 17,272 (therefore more than 92% of the individuals did not report acne). Likewise, the self-reported prevalence of acne in the group of 6,280 individuals who had less than one serving of milk per week was only 6.5%. Therefore, there was a 20% increase in the prevalence of acne in milk drinkers in this study, based solely on memory. More importantly, the self-reported prevalence rates of acne in this study are inconsistent with the well-accepted fact that 70–90% of people are affected by acne at sometime during their teenage years. One who really loves to drink milk could decide to drink two to three glasses and be one of the 15,800 (>92%) individuals who had not developed acne.

Another confusing aspect of that study is that an inverse association was reported between the consumption of milk fat and acne. Most of the hormones present in milk, especially steroids, partition with the milk fat the same way that the fat-soluble vitamins do. These results are bewildering because skim milk would be expected to have fewer hormones due to the processing. It should be expected to have fewer fat-soluble molecules attributing to a lower prevalence than whole milk. No matter how many hormones are left behind in the skim milk, the whole milk should have a higher steroid concentration. Steroids have a structure and partition to fat similar to those of vitamin D. Skim milk is deficient in lipophilic vitamins and by law has to be fortified with the vitamins after removal of the fat. Encouraging intake of vitamin D from other sources could be a mistake as this nutrient plays an important role when calcium is present, which is the case in dairy products and especially milk. The authors also proposed that IGF-I in milk may be present in sufficient quantities to exert biological effects on sebum production. However, whether IGF-I can retain its bioactivity and be protected during processing and in the human digestive tract is uncertain. The intestinal absorption of milk-borne IGF-I is negligible in animals, but no such trials have been demonstrated in humans. Even if hormones are left behind in skim milk, no one has evidence on how much of the various ingested growth factors survive the processing and, most importantly, the human digestive tract.

Indeed, the above studies demonstrated a positive association between milk intake and acne as individuals consuming more milk had a greater prevalence of acne than those with less frequent consumption. Even though absolute values suggested an association of milk with acne shortly thereafter, the AAD (Strauss et al. 2007) was reluctant to implement guidelines based on these data as they were not sufficiently convincing.

The low prevalence rates, memory test, self-assessment, and hormone speculation discussions were not significant enough to drive a recommendation or report on the association of acne to milk. The recommendations were mainly (Strauss et al. 2007) that:

> … a) dietary restriction (either specific foods or food classes) has not been demonstrated to be of benefit in the treatment of acne and b) that there are few clinical studies available in the peer-reviewed literature that directly evaluate the effectiveness of dietary restriction or the consumption of specific foods or food groups to improve acne. These studies failed to support a link between the consumption of chocolate or sugar and acne. Thus, no evidence exists on the role of diet in acne.

An important point is that the dermatologists should not ignore the vast amount of literature on the inverse association of milk or calcium to obesity (Elwood 2005; Zemel and Sun 2008; Zemel 2005; Pereira et al. 2002). Ignoring other studies reporting a positive correlation between lipolysis and calcium and between calcium consumption and weight loss would be a mistake (Zemel 2005; Pereira et al. 2002;

Heiss et al. 2008; Teegarden 2003; Parikh and Yanovski 2003), especially when recommendations and dietary guidelines target children with acne. Certainly any "milk animosity" tendency that has been created should be mediated.

Interestingly, some studies suggested that milk consumption could potentially alter insulin production (Nilsson et al. 2004; Hoyt et al. 2005). Even if milk is responsible for elevated insulin levels, it is noteworthy that higher dairy intake, especially low-fat products, may lower the risk of type 2 diabetes in men and women (Choi et al. 2005; Liu et al. 2006). Although milk has a low glycemic index, the insulin response is comparable to that seen with high-glycemic-index foods. This may be a more important factor in acne development than the ingested hormones and growth factors. It would be good practice to avoid having a cocktail of hormones in our daily diet, but we cannot assume that each person in our society has access or can afford organic or hormone-free dairy products. Indeed, insulin and a high glycemic index are perhaps the two most biochemically and clinically associated factors with acne. There is a relatively adequate amount of research and reports that outline the significance of the insulin pathway on sebaceous biology (Liu et al. 2006; Cappel et al. 2005; Smith et al. 2006; Zouboulis et al. 1998). Recent publications also suggest that PPAR agonists could affect skin and the sebaceous gland (Michalik and Wahli 2007; Zouboulis et al. 2008a; Costello et al. 2007). Especially the PPARγ agonists are well validated as insulin sensitizers, and many of the dietary omega-6 and omega-3 metabolites are PPAR agonists as well (Tontonoz and Spiegelman 2008; Flachs et al. 2009; Tapsell et al. 2009).

Soon after the guidelines of the AAD were published (Strauss et al. 2007), two reports on the association between high glycemic diet and acne were published (Smith et al. 2007a, 2008). Certainly, clinical studies with controlled diets are difficult to perform with ensured compliance, but at least the dermatological community now has clinical data that shed light on the debate over diet and acne. The work of Smith et al. (2007a, 2008) focused on the glycemic load, insulin sensitivity, hormonal mediators, and acne. The investigators reported that foods with a high glycemic index contribute to acne by elevating serum insulin concentrations (which may stimulate sebocyte proliferation and sebum production), suppress SHBG concentrations, and raise androgen concentrations. On the contrary, low-glycemic-index foods increased SHBG and reduced androgen levels, which is important because high SHBG levels were associated with less acne severity.

Acne and insulin sensitivity were assessed after participants followed either a low glycemic load (25% calories from protein, 45% from low-glycemic-index carbohydrates) or a control (typical high glycemic load) diet. Randomly assigned participants (male, $n=43$, 15–25 years old) were enrolled to the dietary intervention or to the control group and were followed for 12 weeks. Blinded dermatologists assessed the number of acne lesions, starting at baseline and then every 4 weeks. Participants on the low-glycemic-load diet experienced greater reductions in total lesion counts and inflammatory lesions than did those in the control group. In addition to in the changes in the acne, volunteers on the low glycemic diet experienced an increase in insulin sensitivity and significant variations in androgen levels (Smith et al. 2007b). A positive correlation was observed between the change in total lesion counts and the change in insulin sensitivity; and SHBG levels correlated negatively with a change in lesion counts.

In all, 31 of the acne patients completed sebum tests as part of the larger 12-week, parallel design dietary intervention trial. At baseline and at the end of the period, follicular sebum outflow and composition of skin surface triglycerides were assessed by instrumental methods. Subjects on the experimental diet demonstrated increases in the ratio of saturated to monounsaturated fatty acids of skin surface triglycerides when compared to controls. That increase was further correlated with acne lesion counts, implicating a possible role of desaturase enzymes in sebaceous lipogenesis and the clinical manifestation of acne.

Recently, Spencer et al. (2009) summarized 27 relevant studies: 21 were observational, and 6 were clinical trials. Bowe et al. (2010) also evaluated the association between diet and acne. They concluded that although there is compelling evidence for the association of acne and high-glycemic-load diets, there is weak evidence for an association between dairy product ingestion and acne.

9.5 Conclusion

The small studies that have been conducted to look at the effect of a low glycemic diet on acne suggest that such a diet could be helpful, although further research is needed to elucidate the role diet may play in acne. The low glycemic diet induces relatively low amounts of insulin to keep blood glucose levels within the normal range, in contrast to a high glycemic diet, which requires more insulin to maintain glucose levels. This could lead to insulin resistance, which in turn causes numerous health problems including high blood pressure, heart disease, obesity, and diabetes.

Because the typical Western diet is similar to the high glycemic diet that often causes insulin resistance, it could potentiate a change in sebum production and therefore inflammation and acne (Smith et al. 2007a, 2008). More research and clinical studies are needed to determine whether a low-glycemic diet could effectively mediate acne or possibly even prevent it.

There are still questions as why not all obese individuals have long-term acne as most people who are obese demonstrate insulin resistance. In addition, if insulin resistance is associated with acne, everyone who suffers from diabetes type 2 would be expected to have acne.

PPARγ agonists (e.g., thiazolidinediones) or dietary fatty acids are known to sensitize cells to insulin in various clinical studies (Tontonoz and Spiegelman 2008; Flachs et al. 2009; Tapsell et al. 2009). Interestingly, PPARγ agonists are also in clinical trials against Alzheimer's disease (AD), which has recently been termed type 3 diabetes (Steen et al. 2005; de la Monte et al. 2006). The human brain uses glucose as its primary fuel, and insulin secreted by the pancreas crosses the blood–brain barrier, reaching neurons and glial cells and potentiating a region-specific effect on glucose metabolism. Glucose homeostasis is critical for energy generation, neuronal maintenance, neurogenesis, neurotransmitter regulation, cell survival, and synaptic plasticity, thereby affecting cognitive function (Hamilton et al. 2007; Cole and Frautschy 2007). There is evidence that demonstrates the efficacy of PPARγ agonists in ameliorating disease-related pathology and improved learning and memory in animal models of AD (de la Monte et al. 2006). Recent clinical trials with PPARγ gamma agonists have shown significant improvement in memory and cognition in AD patients (Landreth et al. 2008). Other studies have shown that omega-3 fatty acids, which are also PPAR ligands, could have a similar effect in AD patients (Fotuhi et al. 2009; Morris et al. 2003).

A recent review referred to the sebaceous gland as the "brain of the skin" and has opened horizons to the newly founded field of dermatoendocrinology (Zouboulis et al. 2008b). This review discussed the links between the endocrine system and sebaceous glands. Apparently, there is an additional link to the relation between the brain and sebaceous cells if we consider that sebaceous and brain cells are dependent on an efficient insulin response for maintenance of proper glucose homeostasis. Therefore, another similarity for sebaceous glands and the brain seems to be the significance that nutrients such as glucose and omega-3 fatty acids have. The latter could positively affect insulin sensitivity and facilitate metabolism of the other most important nutrient, glucose. It is also known that omega-3 fatty acids are preferentially stored in the brain, and research cited previously demonstrated that the precursor of omega-3 s, αLA, targets sebaceous cells (Fu and Sinclair 2000). Perhaps this is another piece of evidence suggesting that the sebaceous gland is the brain of the skin—as their biology is governed by comparable sensitivities in similar nutrients, glucose and omega-3. Does the ectodermic embryonic origin of skin play a role in that similarity? Multiple questions need to be answered by more research in the future.

A high glycemic load seems to be associated with the occurrence of acne, and a recommendation for a low-glycemic load diet cannot harm the affected population. How bad could it be when such a diet, which includes a variety of fruits and vegetables, lean protein, and healthy fats, can also protect against cardiovascular disease, type 2 diabetes, metabolic syndrome, and even obesity? It is noteworthy to reiterate that in the studies by Smith et al. the intervention participants (Smith et al. 2007a, 2008) also lost some weight.

Dermatologists should not ignore nutritional studies, and perhaps nutritionists should understand better the complexity of the skin and sebum production. These specialties should work together to elucidate the "nature of the beast" as it is obvious that much more research and clinical studies are needed to reveal the potential effects of diet or nutrients on acne. We need to understand why people in indigenous societies do not experience acne and, in contrast, the widespread presence of acne throughout modern Western society. Is diet the sole reason, or are other lifestyle and environmental conditions, such as stress, sun exposure, and air pollution, important? To prevent acne by dietary manipulation may not be possible. There are scientifically plausible reasons—beyond the consumers' perceptions—to believe that nutrition can affect acne. To date, research has not proved that diet causes acne. They have provided evidence, though, that diet influences acne to a degree that is still difficult to quantify.

Take Home Messages

- No single dietary element has been shown to be the cause of other noncommunicable diseases. That statement is likely true for acne as well.
- Virtually all Western lifestyle diseases have multifactorial dietary elements that underlie their etiology, and acne is unlikely to be an exception.
- Proper and valid clinical studies are needed to examine the direct relation of diet and acne. So far, only studies that have examined the high glycemic index and load were able to establish a valid relation between diet and acne.

References

Adebamowo CA, Spiegelman D, Danby FW, et al. High school dietary dairy intake and teenage acne. J Am Acad Dermatol. 2005;52:207–14.

Adebamowo CA, Spiegelman D, Berkey CS, et al. Milk consumption and acne in teenaged boys. J Am Acad Dermatol. 2008;58:787–93.

Anderson PC. Foods as the cause of acne. Am Fam Physician. 1971;3(3):102–3.

Bellemère G, Von Stetten O, Oddos T. Retinoic acid increases aquaporin 3 expression in normal human skin. J Invest Dermatol. 2008;128(3):542–8. Epub 2007 Oct 18.

Berger R, Barba A, Fleischer A, Leyden JJ, Lucky A, Pariser D, et al. A double-blinded, randomized, vehicle-controlled, multicenter, parallel-group study to assess the safety and efficacy of tretinoin gel microsphere 0.04% in the treatment of acne vulgaris in adults. Cutis. 2007; 80(2):152–7.

Borland MG, Foreman JE, Girroir EE, Zolfaghari R, Sharma AK, Amin S, et al. Ligand activation of peroxisome proliferator-activated receptorbeta/delta inhibits cell proliferation in human HaCaT keratinocytes. Mol Pharmacol. 2008;74(5):1429–42.

Bowe WP, Joshi SS, Shalita AR. Diet and acne. J Am Acad Dermatol. 2010;63(1):124–41. Epub 2010 Mar 24. Review.

Brown MJ, Ferruzzi MG, Nguyen ML, Cooper DA, Eldridge AL, Schwartz SJ, et al. Carotenoid bioavailability is higher from salads ingested with full-fat than with fatreduced salad dressings as measured with electrochemical detection. Am J Clin Nutr. 2004;80(2):396–403.

Cappel M, Mauger D, Thiboutot D. Correlation between serum levels of insulin-like growth factor 1, dehydroepiandrosterone sulfate, and dihydrotestosterone and acne lesion counts in adult women. Arch Dermatol. 2005;141(3):333–8.

Choi HK, Willett WC, Stampfer MJ, Rimm E, Hu FB. Dairy consumption and risk of type 2 diabetes mellitus in men: a prospective study. Arch Intern Med. 2005;165(9):997–1003.

Cole GM, Frautschy SA. The role of insulin and neurotrophic factor signaling in brain aging and Alzheimer's disease. Exp Gerontol. 2007;42(1-2):10–21. Epub 2006 Oct 17. Review.

Cordain L. Implications for the role of diet in acne. Semin Cutan Med Surg. 2005;24(2):84–91. Review.

Cordain L, Lindeberg S, Hurtado M, Hill K, Eaton SB, Brand-Miller J. Acne vulgaris: a disease of western civilization. Arch Dermatol. 2002;138(12):1584–90.

Costello M, Shrestha B, Eden J, Sjoblom P, Johnson N. Insulin-sensitising drugs versus the combined oral contraceptive pill for hirsutism, acne and risk of diabetes, cardiovascular disease, and endometrial cancer in polycystic ovary syndrome. Cochrane Database Syst Rev. 2007;(1):CD005552. Review.

Danby FW. Diet and acne. Clin Dermatol. 2008;26(1):93–6.

Danby FW. Acne, dairy and cancer: the 5a-P link. Dermatoendocrinol. 2009;1(1):12–6.
de la Monte SM, Tong M, Lester-Coll N, Plater Jr M, Wands JR. Therapeutic rescue of neurodegeneration in experimental type 3 diabetes: relevance to Alzheimer's disease. J Alzheimers Dis. 2006;10(1):89–109. Review.
De Spirt S, Stahl W, Tronnier H, Sies H, Bejot M, Maurette JM, et al. Intervention with flaxseed and borage oil supplements modulates skin condition in women. Br J Nutr. 2009; 101(3):440–5.
Downing D, Strauss J, Pochi P. Changes in skin surface lipid composition induced by severe caloric restriction in man. Am J Clin Nutr. 1972;25:365–7.
Downing DT, Stewart ME, Wertz PW, Strauss JS. Essential fatty acids and acne. J Am Acad Dermatol. 1986;14(2 Pt 1):221–5.
Ekanayake-Mudiyanselage S, Kraemer K, Thiele JJ. Oral supplementation with all-Racand RRR-alpha-tocopherol increases vitamin E levels in human sebum after a latency period of 14–21 days. Ann N Y Acad Sci. 2004;1031:184–94.
Elwood PC. Time to value milk. Int J Epidemiol. 2005;34(5):1160–2.
Fielding JM, Rowley KG, Cooper P, O' Dea K. Increases in plasma lycopene concentration after consumption of tomatoes cooked with olive oil. Asia Pac J Clin Nutr. 2005;14(2):131–6.
Flachs P, Rossmeisl M, Bryhn M, Kopecky J. Cellular and molecular effects of n-3 polyunsaturated fatty acids on adipose tissue biology and metabolism. Clin Sci (Lond). 2009;116(1):1–16. Review.
Fotuhi M, Mohassel P, Yaffe K. Fish consumption, long-chain omega-3 fatty acids and risk of cognitive decline or Alzheimer disease: a complex association. Nat Clin Pract Neurol. 2009;5(3):140–52. Review.
Fu Z, Sinclair AJ. Increased alpha-linolenic acid intake increases tissue alpha-linolenic acid content and apparent oxidation with little effect on tissue docosahexaenoic acid in the guinea pig. Lipids. 2000;35(4):395–400.
Fu Z, Attar-Bashi NM, Sinclair AL. 1–14C-linoleic acid distribution in various tissue lipids of guinea pigs following an oral dose. Lipids. 2001;36(3):255–60.
Fulton Jr JE, Plewig G, Kligman AM. Effect of chocolate on acne vulgaris. JAMA. 1969;210(11):2071–4.
Ge L, Gordon JS, Hsuan C, Stenn K, Prouty SM. Identification of the delta-6 desaturase of human sebaceous glands: expression and enzyme activity. J Invest Dermatol. 2003;120(5):707–14.
Hamilton G, Proitsi P, Jehu L, Morgan A, Williams J, O'Donovan MC, et al. Candidate gene association study of insulin signaling genes and Alzheimer's disease: evidence for SOS2, PCK1, and PPARgamma as susceptibility loci. Am J Med Genet B Neuropsychiatr Genet. 2007;144B(4):508–16.
Harrison EH. Mechanisms of digestion and absorption of dietary vitamin A. Annu Rev Nutr. 2005;25:87–103. Review.
Heiss CJ, Shaw SE, Carothers L. Association of calcium intake and adiposity in postmenopausal women. J Am Coll Nutr. 2008;27(2):260–6.
Horrobin DF. Review. Essential fatty acids in clinical dermatology. J Am Acad Dermatol. 1989;20(6):1045–53.
Hoyt G, Hickey MS, Cordain L. Dissociation of the glycaemic and insulinaemic responses to whole and skimmed milk. Br J Nutr. 2005;93(2):175–7.
http://www.skincarephysicians.com/acnenet/acne_and_diet.html the website is from the AAD American Academy of Dermatology
James MJ, Gibson RA, Cleland LG. Dietary polyunsaturated fatty acids and inflammatory mediator production. Am J Clin Nutr. 2000;71(1 Suppl):343S–8. Review.
Katsambas A, Dessinioti C. New and emerging treatments in dermatology: acne. Dermatol Ther. 2008;21(2):86–95. Review.
Kris-Etherton PM, Taylor DS, Yu-Poth S, Huth P, Moriarty K, Fishell V, et al. Polyunsaturated fatty acids in the food chain in the united states. Am J Clin Nutr. 2000;71(1 Suppl):179S–88. Review.
Landreth G, Jiang Q, Mandrekar S, Heneka M. PPARgamma agonists as therapeutics for the treatment of Alzheimer's disease. Neurotherapeutics. 2008;5(3):481–9. Review.

Liu S, Choi HK, Ford E, Song Y, Klevak A, Buring JE, et al. A prospective study of dairy intake and the risk of type 2 diabetes in women. Diabetes Care. 2006;29(7):1579–84.

Llewellyn A. Variations in the composition of skin surface lipid associated with dietary carbohydrates. Proc Nutr Soc. 1967;26:11.

Logan AC. Omega-3 fatty acids and acne. Arch Dermatol. 2003;139(7):941–2.

MacDonald L. Changes in the fatty acid composition of sebum associated with high carbohydrate diets. Nature. 1964;203:1067–8.

MacDonald L. Dietary carbohydrates and skin lipids. Br J Dermatol. 1967;79:119–21.

Michalik L, Wahli W. Peroxisome proliferator-activated receptors (PPARs) in skin health, repair and disease. Biochim Biophys Acta. 2007;1771(8):991–8. Epub 2007 Feb 24. Review.

Morris MC, Evans DA, Bienias JL, Tangney CC, Bennett DA, Wilson RS, et al. Consumption of fish and n-3 fatty acids and risk of incident Alzheimer disease. Arch Neurol. 2003;60(7):940–6.

Mulokozi G, Hedrén E, Svanberg U. In vitro accessibility and intake of beta-carotene from cooked green leafy vegetables and their estimated contribution to vitamin A requirements. Plant Foods Hum Nutr. 2004;59(1):1–9.

Nicolaides N. Skin lipids: their biochemical uniqueness. Science. 1974;186(4158):19–26. Review.

Nilsson M, Stenberg M, Frid AH, Holst JJ, Björck IM. Glycemia and insulinemia in healthy subjects after lactose-equivalent meals of milk and other food proteins: the role of plasma amino acids and incretins. Am J Clin Nutr. 2004;80(5):1246–53.

Orfanos CE, Zouboulis CC, Almond-Roesler B, Geilen CC. Current use and future potential role of retinoids in dermatology. Drugs. 1997;53(3):358–88. Review.

Pappas A, Anthonavage M, Gordon JS. Metabolic fate and selective utilization of major fatty acids in human sebaceous gland. J Invest Dermatol. 2002;118(1):164–71.

Parikh SJ, Yanovski JA. Calcium intake and adiposity. Am J Clin Nutr. 2003;77(2):281–7. Review.

Pereira MA, Jacobs Jr DR, Van Horn L, Slattery ML, Kartashov AI, Ludwig DS. Dairy consumption, obesity, and the insulin resistance syndrome in young adults: the CARDIA study. JAMA. 2002;287(16):2081–9.

Pochi P, Downing D, Strauss J. Sebaceous gland response in man to prolonged total caloric deprivation. J Invest Dermatol. 1970;55:303–9.

Popadic S, Ramic Z, Medenica L, Mostarica Stojkovic M, Trajković V, Popadic D. Antiproliferative effect of vitamin A and D analogues on adult human keratinocytes in vitro. Skin Pharmacol Physiol. 2008;21(4):227–34.

Reichrath J, Lehmann B, Carlberg C, Varani J, Zouboulis CC. Vitamins as hormones. Horm Metab Res. 2007;39(2):71–84. Review.

Ribaya-Mercado JD. Influence of dietary fat on beta-carotene absorption and bioconversion into vitamin A. Nutr Rev. 2002;60(4):104–10. Review.

Rubin MG, Kim K, Logan AC. Acne vulgaris, mental health and omega-3 fatty acids: a report of cases. Lipids Health Dis. 2008;7:36.

Russell RM. Vitamin and trace mineral deficiency and excess. In: Braunwald E, Fauci A, Kasper D, et al., editors. Harrison's principles of internal medicine, vol. 1. 15th ed. New York: McGraw-Hill; 2001. p. 465–6.

Smith TM, Cong Z, Gilliland KL, Clawson GA, Thiboutot DM. Insulin-like growth factor-1 induces lipid production in human SEB-1 sebocytes via sterol response elementbinding protein-1. J Invest Dermatol. 2006;126(6):1226–32.

Smith RN, Mann NJ, Braue A, et al. A low-glycemic-load diet improves symptoms in acne vulgaris patients: a randomized controlled trial. Am J Clin Nutr. 2007a;86:107–15.

Smith RN, Mann NJ, Braue A, et al. The effect of a high-protein, low glycemic-load diet versus a conventional, high glycemic-load diet on biochemical parameters associated with acne vulgaris: a randomized, investigator-masked, controlled trial. J Am Acad Dermatol. 2007b;57:247–56.

Smith RN, Braue A, Varigos GA, et al. The effect of a low glycemic load diet on acne vulgaris and the fatty acid composition of skin surface triglycerides. J Dermatol Sci. 2008;50:41–52.

Spencer EH, Ferdowsian HR, Barnard ND. Diet and acne: a review of the evidence. Int J Dermatol. 2009;48(4):339–47. review.

Steen E, Terry BM, Rivera EJ, Cannon JL, Neely TR, Tavares R, et al. Impaired insulin and insulin-like growth factor expression and signaling mechanisms in Alzheimer's disease–is this type 3 diabetes? J Alzheimers Dis. 2005;7(1):63–80.

Strauss JS, Krowchuk DP, Leyden JJ, Lucky AW, Shalita AR, Siegfried EC, et al. American Academy of Dermatology/American Academy of Dermatology Association. Guidelines of care for acne vulgaris management. J Am Acad Dermatol. 2007;56(4):651–63. Epub 2007 Feb 5. Review.

Tapsell LC, Batterham MJ, Teuss G, Tan SY, Dalton S, Quick CJ, et al. Long-term effects of increased dietary polyunsaturated fat from walnuts on metabolic parameters in type II diabetes. Eur J Clin Nutr. 2009;63(8):1008–15.

Teegarden D. Calcium intake and reduction in weight or fat mass. J Nutr. 2003;133(1):249S–51. Review.

Thiboutot DM. Overview of acne and its treatment. Cutis. 2008;81:3–7.

Thiele JJ, Weber SU, Packer L. Sebaceous gland secretion is a major physiologic route of vitamin E delivery to skin. J Invest Dermatol. 1999;113(6):1006–10.

Tontonoz P, Spiegelman BM. Fat and beyond: the diverse biology of PPARgamma. Annu Rev Biochem. 2008;77:289–312.

Trebble T, Arden NK, Stroud MA, Wootton SA, Burdge GC, Miles EA, et al. Inhibition of tumour necrosis factor-alpha and interleukin 6 production by mononuclear cells following dietary fish-oil supplementation in healthy men and response to antioxidant co-supplementation. Br J Nutr. 2003;90(2):405–12.

Treloar V. Comment on the commentary: diet and acne. J Am Acad Dermatol. 2008;59(3):534–5.

Treloar V, Logan AC, Danby FW, Cordain L, Mann NJ. Comment on acne and glycemic index. J Am Acad Dermatol. 2008;58(1):175–7.

Tsukada M, Schröder M, Roos TC, Chandraratna RA, Reichert U, Merk HF, et al. 13-cis retinoic acid exerts its specific activity on human sebocytes through selective intracellular isomerization to all-trans retinoic acid and binding to retinoid acid receptors. J Invest Dermatol. 2000;115(2):321–7.

Unlu NZ, Bohn T, Clinton SK, Schwartz SJ. Carotenoid absorption from salad and salsa by humans is enhanced by the addition of avocado or avocado oil. J Nutr. 2005;135(3):431–6.

Vahlquist A. What are natural retinoids? Dermatology. 1999;199 Suppl 1:3–11. Review.

Webster G. Commentary: diet and acne. J Am Acad Dermatol. 2008;58(5):794–5.

Webster G, Cargill DI, Quiring J, Vogelson CT, Slade HB. A combined analysis of 2 randomized clinical studies of tretinoin gel 0.05% for the treatment of acne. Cutis. 2009;83(3):146–54.

Wolf R, Matz H, Orion E. Acne and diet. Clin Dermatol. 2004;22(5):387–93. Review.

Zemel MB. The role of dairy foods in weight management. J Am Coll Nutr. 2005;24 (6 Suppl):537S–46. Review.

Zemel MB, Sun X. Dietary calcium and dairy products modulate oxidative and inflammatory stress in mice and humans. J Nutr. 2008;138(6):1047–52.

Zouboulis CC. Is acne vulgaris a genuine inflammatory disease? Dermatology. 2001;203(4):277–9.

Zouboulis CC. Acne and sebaceous gland function. Clin Dermatol. 2004;22:360–6.

Zouboulis CC, Xia L, Akamatsu H, Seltmann H, Fritsch M, Hornemann S, et al. The human sebocyte culture model provides new insights into development and management of seborrhoea and acne. Dermatology. 1998;196(1):21–31. Review.

Zouboulis CC, Nestoris S, Adler YD, Orth M, Orfanos CE, Picardo M, et al. A new concept for acne therapy: a pilot study with zileuton, an oral 5-lipoxygenase inhibitor. Arch Dermatol. 2003;139(5):668–70.

Zouboulis ChC, Saborowski A, Boschnakow A. Zileuton, an oral 5-lipoxygenase inhibitor, directly reduces sebum production. Dermatology. 2005;210(1):36–8.

Zouboulis CC, Schagen S, Alestas T. The sebocyte culture: a model to study the pathophysiology of the sebaceous gland in sebostasis, seborrhoea and acne. Arch Dermatol Res. 2008a;300(8):397–413.

Zouboulis CC, Baron JM, Böhm M, Kippenberger S, Kurzen H, Reichrath J, et al. Frontiers in sebaceous gland biology and pathology. Exp Dermatol. 2008b;17(6):542–51. Review.

Chapter 10
Glycemic Load and Acne

Robyn Smith and Neil Mann

Core Messages

- Acne affects 80–90% of individuals during puberty, which is a period associated with a normal decline in insulin sensitivity (Bloch et al. 1987). Hyperinsulinemia stemming from insulin resistance may play a key role in activating a hormonal milieu conducive for acne development.
- Acne remains relatively unknown among non-Westernized societies existing on low-glycemic-load diets, and prevalence rates increase when a previously unaffected society is exposed to refined, high-glycemic-index carbohydrates.
- Low-glycemic-load diets may represent a unique dietary strategy for alleviating acne symptoms through improving insulin metabolism.

10.1 Introduction

Acne has long been thought to be associated with the consumption of certain foods. During the early 1900s, dermatology textbooks commonly noted that diets high in carbohydrates and sweets tended to make acne worse, with chocolate thought to be the most offending factor (Wise and Sulzberger 1933). Elimination diets were reported to have moderate success (Cormia 1940; White 1934). However, there was little agreement among dermatologists about what foods should be avoided. Allergenic skin tests were unable to identify the culprit foods (Cormia 1940; White 1934), suggesting that foods may aggravate acne through an indeterminate mechanism.

R. Smith (✉)
Baker IDI Heart and Diabetes Institute, PO Box 6492,
St Kilda Road Central, Melbourne, VIC 8008, Australia
e-mail: robyn.smith@bakeridi.edu.au

A. Pappas (ed.), *Nutrition and Skin: Lessons for Anti-Aging, Beauty and Healthy Skin*,
DOI 10.1007/978-1-4419-7967-4_10, © Springer Science+Business Media, LLC 2011

Consequently, dermatologists had to rely on their own clinical experience, and there was no unified approach to deal with the condition. The diet and acne connection finally fell from favor in 1969 when a study found no difference in acne after the daily consumption of a chocolate versus a placebo bar containing equivalent amounts of fat and sugar (Fulton et al. 1969). Although this study was criticized at the time for several design flaws (Mackie and Mackie 1974), it brought a long awaited end to the trial-and-error dietary practices in the management of acne.

It is important to consider that at the time of the chocolate study the main nutritional challenge was the prevention of dietary deficiencies. Consequently, few could comprehend the notion of food causing disease in the absence of any real metabolic or nutritional deficiency. From 1958 to 1979, dietary advice was given in terms of minimum daily requirements, and the recommendation was to eat more than the dietary guidelines suggested to satisfy energy and nutrient needs (Welsh et al. 1993). The notion to "eat more" to prevent deficiency was in clear opposition to the idea of avoiding certain foods to prevent acne. The authors of the chocolate study summarized the consideration of the time that "if a food can really alter a disease … [this] finding would set into motion a wholesale attack on the effects of foods on normal physiologic functions." However, in 1969 scientists were just beginning to become aware of the role of diet in the etiology of chronic disease states such as cardiovascular disease and diabetes. By the late 1970s, dietary advice had shifted from preventing dietary deficiencies to recommendations aimed at avoiding excessive consumption of food components (e.g., fat, saturated fat, cholesterol, sugar, salt, alcohol). Public awareness of the cause-and-effect relationship of the role of food in general health has since caused an explosion in scientific and mainstream literature as well as the development of new nutritional concepts such as the glycemic index. This megatrend has fueled an explosion in nutritional research and a new understanding of how the food we eat relates to health or particular pathological processes.

Since the chocolate study, there has been a logarithmic progression in the science of nutrition that has generally gone unnoticed in the dermatology field. Over the past three decades, considerable focus has been given to the role of dietary carbohydrate in the rising tide of obesity and the cluster of metabolic disturbances (hyperinsulinemia, insulin resistance, hyperglycemia, hypertension, dyslipidemia) now termed the metabolic syndrome. Because the metabolic syndrome is considered to be a condition of disordered insulin metabolism, there is merit in evaluating foods based on their rate of absorption and their effect on blood glucose and insulin concentrations. Consequently, the glycemic index (GI) was introduced to quantify the blood glucose-raising potential for a given sample of food containing 50 g of available carbohydrate (Jenkins et al. 1981). By definition, high-GI carbohydrates are rapidly digested, producing rapid elevations in blood glucose and increasing insulin demand. In contrast, low-GI carbohydrates are slowly digested and absorbed, and they elicit a low insulin response during the postprandial period. It is thought that the abundance of low-GI foods and the absence of refined, high-GI carbohydrates in traditional cultures together may play a protective role against Western diseases (Colagiuri and Brand-Miller 2002). The evidence to date suggests that low-GI diets are associated with higher high-density lipoprotein (HDL)-cholesterol levels and with a lower risk of developing cardiovascular disease and diabetes (Liu 2002).

Furthermore, dietary intervention trials suggest that low-GI diets may increase satiety and facilitate weight loss when compared to high-GI diets (Ludwig 2000). Therefore, the glycemic index has been a useful nutritional concept, providing new insights into the relation of food and Western diseases.

This chapter reconsiders the diet and acne connection and opens a new era for understanding the effects of nutrition on skin health. According to early scientific principles of medicine, to test the hypothesis that a dietary component is implicated in disease, one should demonstrate that (1) the diet of persons with disease is significantly different from those without the disease; (2) the signs and symptoms should be known to be or plausibly suspected of being caused by the dietary imbalance; and (3) correction of the dietary imbalance should result in alleviation of the signs and symptoms. Although these fundamental principles should be simple to apply, the major limitation in 1969 was poor understanding of the cause and exacerbation of acne. A better understanding of the disease at the biochemical level has helped elucidate a possible role of nutritionally related factors in acne etiology.

10.2 Acne Pathophysiology: Could Acne Be a Metabolic Disease?

Acne is considered as a disease of adolescence, affecting 80%–90% of individuals aged 12–15 years (Lucky et al. 1991). Clinical observation indicates that this condition can also affect prepubescent children and adults. Acne often begins during adrenarche (8–12 years), and incidence rates increase with pubertal maturation. During adolescence, the disease is more common and severe in boys, possibly reflecting an effect of androgens on sebum production (Stathakis et al. 1997). Acne incidence declines after 18 years of age, but a considerable number of men and women aged 20–40 years continue to be affected. A review of the data over recent years suggests an increasing prevalence of acne in people over the age of 25, particularly women (Goulden et al. 1997). It is unknown why acne tends to be chronic for a subset of adult women, but severity is reported to be influenced by factors that reflect hormonal fluctuations, including menstrual cycles, pregnancy, and menopause (Shaw and White 2001; Thiboutot and Lookingbill 1995).

Examination of the life course of acne may provide a physiological framework on which we can examine the role of diet-related factors in acne development. Acne generally begins when androgens of either adrenal or gonadal origin increase, stimulating sebum production. The most important androgen is testosterone, which may be locally converted to the more active dihydrotestosterone by 5α-reductase. However, acne severity and incidence does not correlate well with testosterone levels, suggesting that the hormonal control of acne is complex and may involve interplay of other factors. Acne has shown to correlate better with the proportion of testosterone to sex hormone binding globulin (SHBG), an indicator of testosterone bioavailability. Other biological factors, such as insulin and insulin-like growth factor (IGF)-I, can also stimulate sebum production and growth of keratinocytes in cell cultures (Deplewski and Rosenfield 2000; Eming et al. 1996).

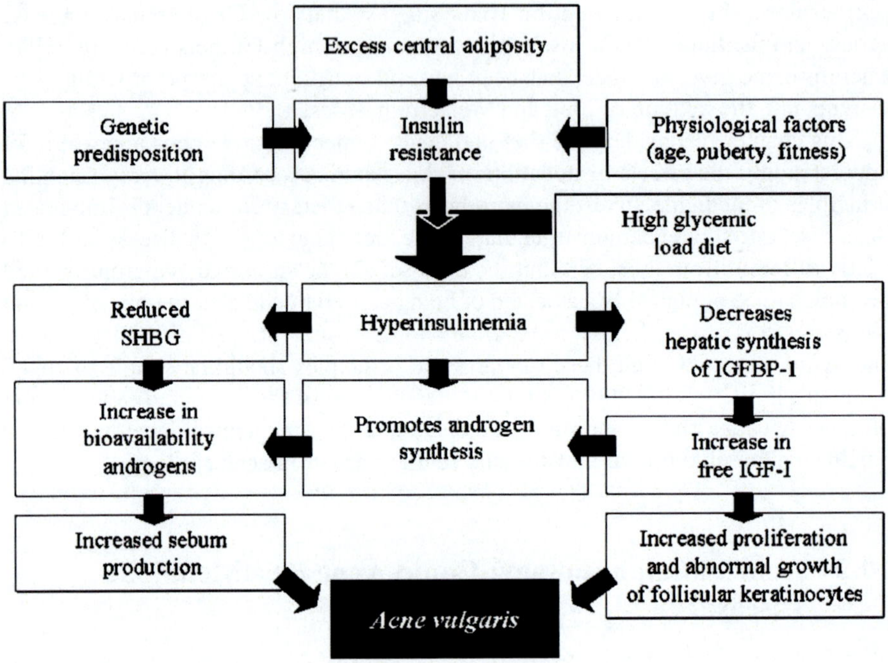

Fig. 10.1 Mechanism of a high glycemic load diet promoting acne in insulin-resistant states. *SHBG* sex hormone binding globulin, *IGFBP-1* insulin-like growth factor binding protein-1, *IGF-I* insulin-like growth factor-I

Clinically, elevated insulin levels have been described in women with persistent adult acne (Aizawa and Niimura 1996), and significantly higher IGF-I levels have been described in women with acne compared with controls (Deplewski and Rosenfield 1999; Aizawa and Niimura 1995; Cappel et al. 2005). These physiological traits may influence one or more of the pathogenic processes involved in acne development, including (1) increased sebum production, (2) hyperproliferation and differentiation of follicular keratinocytes, (3) microbial colonization by *Propionbacterium acnes*, and (4) inflammation.

Among the factors associated with the clinical presentation of acne, insulin may play a key role in activating a hormonal milieu conducive for acne development (Fig. 10.1). Insulin has been shown to augment the growth-promoting effects of IGF-I by decreasing levels of its binding protein, IGF-binding protein 1 (IGFBP-1) (Powell et al. 1991). Both insulin and IGF-I can stimulate adrenal androgen synthesis and gonadal testosterone production through effects on steroidogenic enzymes and gonadotrophin-releasing hormone secretion (Willis et al. 1996). Insulin and IGF-I can also inhibit hepatic SHBG production (Singh et al. 1990), increasing the concentration of androgens that are bioavailable. Therefore, in addition to the direct effects of insulin on pathogenic processes (i.e., sebum production, keratinocyte growth), insulin has the potential to indirectly affect acne through shifts in endocrine systems.

Support for a role of insulin in acne development can be found in the pattern of acne prevalence. Acne clinically presents during puberty, which is a transient period of normal insulin resistance (Bloch et al. 1987; Moran et al. 1999). Even before puberty, hyperinsulinemia is found to be predictive of acne incidence in later years (Miller et al. 1996; Ibáñez et al. 1997). Acne incidence follows the rise and fall of pubertal insulin resistance more closely than the change in androgen levels, as androgen concentrations remain elevated following acne regression during the late teenage years. Fluctuations in the incidence of acne throughout the normal life cycle appears to coincide with changes in insulin sensitivity, with insulin resistance observed also during pregnancy, menses, and menopause (Homko et al. 2001; Pulido and Salazar 1999; Godsland et al. 1995). Perhaps the strongest evidence for an association of acne and insulin resistance comes from the fact that acne is a common feature of women with polycystic ovary syndrome (PCOS), a condition characterized by hyperinsulinemia and hyperandrogenism. Clinical observation suggests that insulin resistance is the underlying feature in PCOS, as it generally precedes and gives rise to hyperandrogenism (Dunaif et al. 1989). Furthermore, reducing insulin secretion and/or increasing insulin sensitivity with pharmacological interventions (i.e., acarbose and metformin, respectively) decreases the severity of acne symptoms in individuals with PCOS (Ciotta et al. 2001; Kazerooni and Dehghan-Kooshkghazi 2003; Kolodziejczyk et al. 2000).

If insulin resistance and hyperinsulinemia are suspected to play a role in acne development, it suggests that obese hyperinsulinemic adults should also exhibit signs of acne. The absence of acne in older, obese adults may be because, during puberty, both insulin resistance and acne are intrinsically linked to the growth hormone/IGF-I axis (Deplewski and Rosenfield 1999; Moran et al. 2002a). Acne typically presents prior to puberty when growth hormone (GH) and IGF-I levels begin to rise; and usually it resolves by the third decade as GH and IGF-I levels decline. In obese hyperinsulinemic adults, however, there is a negative association between adiposity and GH levels (Pijl et al. 2001; Luque and Kineman 2006). Obese individuals not only have a low basal GH output, they exhibit blunted responses to all recognized GH stimuli, including fasting, acute exercise, GH-releasing hormone, and insulin tolerance tests (Luque and Kineman 2006; Williams et al. 1984; Qu et al. 2004). Some individuals continue to exhibit high IGF-I levels beyond puberty, and elevated IGF-I levels have been found to be a precipitating factor in persistent adult acne (Aizawa and Niimura 1995). This suggests that the status of the GH/IGF-I axis may be an important underlying factor in acne pathogenesis.

10.3 Secular Trends of Advancing Pubertal Age and Acne: Evidence for a Role of Diet?

The reasons for the change in insulin sensitivity at the various hormonal stages of life (i.e., pregnancy, puberty, menses, menopause) are unknown. It has been suggested that insulin resistance of puberty may relate more to changes in growth hormone release than changes in body mass (Moran et al. 2002b). Interestingly, in Western countries, the mean age of menarche has fallen from 16 to 13 years since the beginning of the century, which suggests an earlier onset of pubertal insulin resistance and hyperinsulinemia. Although the insulin resistance of puberty is thought to be relatively benign, exposure to hyperinsulinemia at a younger age is predicative of the later development of states of androgen excess: PCOS, acne, hirsutism (Miller et al. 1996). Western nutrition is generally assumed to be responsible for the secular trend to an ever-earlier onset of puberty. Although the nutritional factors responsible remain to be identified, possible metabolic cues include degree of fatness ("critical fatness"), glucose availability, and IGF-I and leptin levels (Foster and Nagatani 1999).

Epidemiological observations have provided us with some compelling evidence for a role of Western dietary factors in acne development and may provide us with some insight into the role of nutrition in advancing pubertal development. A recent observational report described the low incidence of acne in non-Westernized societies, where the mean age of menarche was 16 years (Cordain et al. 2002). The authors implicated diet—mainly the absence of high-GI foods—for the low rates of acne in these societies. This is in support of earlier observations during the 1970s of the emergence of acne in Eskimos groups following the introduction of Western foods (Schaefer 1971; Bendiner 1974). The higher rates of acne in Eskimo groups paralleled the increase in annual per-capita consumption of refined sugar and flour, and the per-capita consumption of protein from animal sources showed an inverse relation. Even though the Eskimo's traditional hunter-gatherer diet was very low in carbohydrate, there was a relatively high intake of carbohydrate following the introduction of agriculture by Russian settlers some 70–100 years ago (Schaefer 1970). However, these carbohydrates (e.g., barley, buckwheat, cabbage, potatoes) had a relatively low glycemic index (Table 10.1) and did not replace animal protein as the main source of energy. Only since the relatively recent exposure to refined high-GI carbohydrates have the Eskimos demonstrated faster growth (increased final height), earlier puberty, and dramatic increases in the incidence of obesity, diabetes, and heart disease.

The traditional diets of acne-free populations characteristically have a low glycemic load and therefore produce only modest postprandial rises in plasma glucose and insulin. As the glycemic index can only be used to compare foods of equal carbohydrate content, the glycemic load was later developed to characterize the glycemic effect of whole meals and diets on the basis of the rate of glucose appearance and the quantity of carbohydrate consumed (glycemic index × carbohydrate content). At present, there are relatively few data available on the classification of foods according to their insulin response, although the correlation between glycemic and insulinemic responses is reported to be high ($r=0.74$ and 0.90, respectively) (Holt et al. 1997; Bornet et al. 1987). Table 10.1 illustrates that food staples of traditional cultures elicit lower glycemic and insulinemic responses than Western dietary staples. When applying the glycemic load concept to whole diets, the glycemic load may be reduced by decreasing total carbohydrate intake or by selecting foods using the glycemic index concept. Traditionally, the Eskimo diet

Table 10.1 Glycemic and insulinemic responses to foods commonly consumed in traditional (acne-free) cultures and Westernized societies

Food	Serving weight (g)	Glycemic index[a] (classification)	Glycemic load	Insulin response per gram serving weight[b]
Foods commonly consumed in traditional cultures				
White rice	150	83 (high)	36	40
Barley	150	27 (low)	11	–
Buckwheat	150	51 (low)	15	–
Lentils, cooked	150	28 (low)	5	37
Eggs	100	0	0	30
Beef	230	0	0	50
Fish	110	0	0	28
Cabbage	100	0	0	–
Sweet potatoes	150	44	11	–
Potatoes, russet	150	66 (mod)	13	38
Apples	120	40 (low)	6	20
Oranges	120	40 (low)	4	15
Bananas	120	62 (mod)	16	45
Black grapes	120	59 (mod)	11	31
Foods commonly consumed in Western societies				
Potato crisps	50	60 (mod)	12	186
French fries, oven baked	150	64 (mod)	21	82
Vanilla ice cream	50	57 (mod)	6	103
Strawberry fruit yoghurt	200	30 (low)	5	65
Mars bar	60	62 (high)	25	309
Doughnuts with cinnamon sugar	50	75 (high)	15	191
Croissants	57	67 (high)	17	215
Chocolate cake with frosting	111	38 (low)	20	223
Water crackers	25	78 (high)	14	253
Chocolate chip biscuits	50	47 (low)	17	298

(continued)

Table 10.1 (continued)

Food	Serving weight (g)	Glycemic index[a] (classification)	Glycemic load	Insulin response per gram serving weight[b]
White bread	30	71 (high)	10	137
Wholemeal bread	30	72 (high)	9	111
Special K cereal	30	54 (low)	11	47
Cornflakes cereal	30	77 (high)	19	52
Honey smacks® cereal	30	71 (high)	16	53

[a]GI and GL values have been sourced from GI tables (Sydney University 2010; Foster-Powell et al. 2002), which used glucose as the reference food. Glycemic index classification: low < 55; moderate 56–69; high > 70

[b]Values for the insulin response per gram serving weight was sourced from Holt et al. (1997)

would have been low in glycemic load due to low intakes of carbohydrate and the consumption of low-GI foods. Acne became a problem in these societies only when the adolescents began to consume high-GI carbohydrates (i.e., sweet biscuits, potato crisps, soft drinks, confectionary) in large quantities (Schaefer 1971; Bendiner 1974). Consequently, reducing the dietary glycemic load may represent a unique dietary strategy to alleviate acne via a reduction in hyperinsulinemia and its hormonal sequelae.

10.4 Clinical Evidence of a Therapeutic Effect of Low Glycemic Load Diets in Acne Vulgaris

A recent randomized controlled trial found that a low glycemic load (GL) diet that mimics the diets of acne-free populations may alleviate acne symptoms and hormonal markers of acne. In a 12-week study, 43 young male subjects (age 15–25 years) with mild to moderate acne consumed either a conventional high-GL diet or a low-GL diet and had their acne assessed every 4 weeks (Smith et al. 2007a, b). The experimental low-GL diet was achieved through a reduction in carbohydrate intake and through selection of low-GI foods. After 12 weeks, study participants on the lowGL diet demonstrated a 20% greater reduction in acne lesion counts than the participants on the high-GL diet. The lessening of acne severity can be explained by improvements in metabolic-endocrine parameters. When compared to controls, participants on the low-GL diet demonstrated significant improvements in insulin sensitivity and hormonal markers of acne (increases in SHBG and IGFBP-1, suggesting decreased bioavailability of testosterone and IGF-I) (Smith et al. 2007b). These changes may also relate to the modest weight loss (2.5 kg) that occurred with the reduction in dietary glycemic load. Post hoc analyses revealed that the effect of the low-GL diet on acne and certain endocrine parameters was lost after statistically adjusting for weight loss. This suggests that weight loss mediated the reduction in insulin resistance and its associated hyperinsulinemia, which may be important in the clinical regression of acne.

At present, it remains uncertain whether diet can alleviate acne without weight loss. In theory, hyperinsulinemia can be reduced through improvements in the metabolic state of insulin resistance and/or reducing postprandial hyperglycemia. Interestingly, another study found no difference in the global assessment of acne in weight-maintained individuals following the consumption of diets that had a high and a low glycemic index (Reynolds et al. 2010, personal communication). Although this suggests that weight loss may be responsible for the clinical regression in the earlier study, it should be noted that differences in the nature of the dietary interventions (reduced dietary glycemic index versus reduced dietary glycemic load) and the assessment of acne (acne lesion counts versus global assessment of acne) may also account for the different study outcomes.

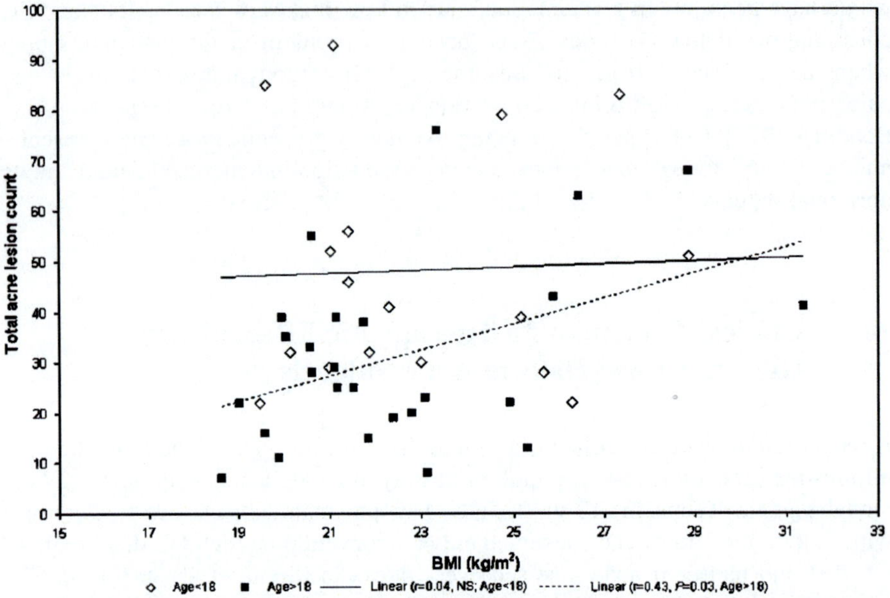

Fig. 10.2 Age-dependent association of acne severity and body mass index (*BMI*). Bivariate analysis was performed with a two-tailed Pearson's correlation ($n=43$)

Few studies have reported an association of body weight and the incidence of acne. The U.S. National Health Survey of 1966–1970 found that dietary excess (as reported by the parent) was significantly associated with a greater degree of acne prevalence, particularly in young boys (US Department of Health Education and Welfare 1976). A survey of 2,720 soldiers demonstrated that adult men with acne were significantly heavier (5.6 kg) than adult men without acne. However, this association was found to be dependent on age, as weight was not associated with age in adolescents aged 15–19 years. Interestingly, a similar age-dependent association between acne and obesity was found in the low-GL study. This study found that acne lesion counts correlated with body mass index (BMI) in men ≥18 years of age, but this association was not true for boys <18 years of age (Fig. 10.2). The reason for the age-dependent association is unknown, but it is possible that acne during adolescence may relate more to physiological changes in insulin sensitivity (via a weight-independent mechanism), whereas adult acne may be more pathological in origin (obesity-related insulin resistance).

These clinical observations may provide a foundation for future dietary recommendations in the management of acne. There is already scope for diet as a treatment option, as acne patients often treat themselves with over-the-counter therapies. Diet therapy may be used alone or in conjunction with conventional acne therapies in cases of mild to moderate acne. However, as severe nodulocystic acne can be painful, disfiguring, and leave permanent scars, it is recommended that patients seek optimal treatment from a specialist physician. When considering diets for adolescents, there are two eating behaviors that also require special consideration: eating that leads to obesity and the disturbed psychiatric conditions of anorexia and bulimia. Like acne, obesity can have significant negative psychological consequences, including low self-esteem, social inhibition, depression, and anxiety. Low-GL diets may present a useful strategy for individuals seeking to lose weight and at the same time prevent or lessen the severity of acne. However, diet should not be considered as a strategy for psychologically vulnerable individuals with a preexisting eating disorder, as it may increase the food-associated anxiety and the preoccupation with food.

There is now also compelling evidence from clinical and epidemiological studies to suggest a potential role of diet in acne development. In 1969, the authors of the chocolate study stated that "it would be remarkable if skin functions were easily influenced by the vagaries of the diverse diets which have evolved in human populations." During the four-decade scientific vacuum since the chocolate study, much has been learned from the diets of non-Westernized societies. Dietary intervention trials suggest that low-GL diets can alleviate acne symptoms, possibly through improving insulin metabolism and decreasing the bioavailability of testosterone and IGF-I. These endocrine changes may influence the desquamation of follicular keratinocytes and sebum production, two primary factors involved in the development of an acne lesion. It remains to be objectively determined whether weight loss is the principal factor in the clinical alleviation of acne.

Take Home Messages

- A recent randomized controlled trial found that a low-glycemic-load diet that mimics the diets of acne-free populations improves acne symptoms when compared to a high glycemic load diet.
- Low-glycemic-load diets may alleviate acne by decreasing insulin demand and influencing mediators such as sex-hormone binding globulin, insulin-like growth factor-I, and insulin-like growth factor-binding proteins. Together, these may affect several aspects of the disease process, including sebum production and desquamation of follicular keratinocytes.
- It remains to be objectively determined whether weight loss associated with low-glycemic-load diets is the principal factor for the alleviation of acne.

References

Aizawa H, Niimura M. Elevated serum insulin-like growth factor-I (IGF-I) levels in women with postadolescent acne. J Dermatol. 1995;22(4):249–52.

Aizawa H, Niimura M. Mild insulin resistance during oral glucose tolerance test (OGTT) in women with acne. J Dermatol. 1996;23:526–9.

Bendiner E. Disastrous trade-off: Eskimo health for white civilization. Hosp Pract. 1974;9:156–89.
Bloch C, Clemons P, Sperling M. Puberty decreases insulin sensitivity. J Pediatr. 1987;110:481–7.
Bornet F et al. Insulinemic and glycemic indexes of six starch-rich foods taken alone and in a mixed meal by type 2 diabetics. Am J Clin Nutr. 1987;45:588–95.
Cappel M, Mauger D, Thiboutot D. Correlation between serum levels of insulin-like growth factor 1, dehydroepidandrosterone sulfate, and dihydrotestosterone and acne lesion counts in adult women. Arch Dermatol. 2005;141:333–8.
Ciotta L et al. Clinical, endocrine and metabolic effects of acarbose, an a-glucosidase inhibitor, in PCOS patients with increased insulin response and normal glucose tolerance. Hum Reprod. 2001;16(10):2066–72.
Colagiuri S, Brand-Miller J. The 'carnivore connection' – evolutionary aspects of insulin resistance. Eur J Clin Nutr. 2002;56:S30–5.
Cordain L et al. Acne vulgaris – a disease of Western civilization. Arch Dermatol. 2002;138:1584–90.
Cormia F. Food sensitivity as a factor in the etiology of acne vulgaris. J Allergy. 1940;12:34–41.
Deplewski D, Rosenfield R. Growth hormone and insulin-like growth factors have different effects on sebaceous cell growth and differentiation. Endocrinology. 1999;140:4089–94.
Deplewski D, Rosenfield R. Role of hormones in pilosebaceous unit development. Endocr Rev. 2000;21:363–92.
Dunaif A et al. Profound peripheral insulin resistance, independent of obesity in polycystic ovary syndrome. Diabetes. 1989;38:1165–74.
Eming S et al. Targeted expression of insulin-like growth factor to human keratinocytes: modification of the autocrine control of keratinocyte proliferation. J Invest Dermatol. 1996;107:113–20.
Foster D, Nagatani S. Physiological perspectives on leptin as a regulator of reproduction: role in timing puberty. Biol Reprod. 1999;60:205–15.
Foster-Powell K, Holt S, Brand-Miller J. International table of glycemic index and glycemic load values. Am J Clin Nutr. 2002;76:5–56.
Fulton J, Plewig G, Kligman A. Effect of chocolate on acne vulgaris. JAMA. 1969;210(11):2071–4.
Godsland I et al. The insulin resistance syndrome in postmenopausal women with cardiological syndrome X. Br Heart J. 1995;74:47–52.
Goulden V, Clarke SM, Cunliffe WJ. Post adolescent acne: a review of clinical features. Br J Dermatol. 1997;136(1):66–70.
Holt SH, Miller JC, Petocz P. An insulin index of foods: the insulin demand generated by 1000 kJ portions of common foods. Am J Clin Nutr. 1997;66:1264–76.
Homko C et al. Insulin secretion during and after pregnancy in patients with gestational diabetes mellitus. J Clin Endocrinol Metab. 2001;86:568–73.
Ibáñez L et al. Hyperinsulinemia and decreased insulin-like growth factor-binding protein-1 are common features in prepubertal and pubertal girls with a history of premature pubarche. J Clin Endocrinol Metab. 1997;82:2283–8.
Jenkins D, Wolever T, Taylor R. Glycaemic index of foods: a physiological basis for carbohydrate exchange. Am J Clin Nutr. 1981;134:362–6.
Kazerooni T, Dehghan-Kooshkghazi M. Effect of metformin therapy on hyperandrogenism in women with polycystic ovary syndrome. Gynecol Endocrinol. 2003;17:51–6.
Kolodziejczyk B et al. Metformin therapy decreases hyperandrogenism and hyperinsulinemia in women with polycystic ovary syndrome. Fertil Steril. 2000;73:1149–54.
Liu S. Intake of refined carbohydrates and whole grains in relation to risk of type 2 diabetes mellitus and coronary heart disease. J Am Coll Nutr. 2002;21:298–306.
Lucky A et al. Acne vulgaris in early adolescent boys: correlations with pubertal maturation and age. Arch Dermatol. 1991;127:210–6.
Ludwig D. Dietary glycemic index and obesity. J Nutr. 2000;130 Suppl 2:280S–3.
Luque R, Kineman R. Impact of obesity on the growth hormone axis: evidence for a direct inhibitory effect of hyperinsulinemia on pituitary function. Endocrinology. 2006;147:2754–63.
Mackie B, Mackie L. Chocolate and acne. Aust J Dermatol. 1974;15:103–9.
Miller D, Emans S, Kohane I. Follow-up study of adolescent girls with a history of premature pubarche. J Adolesc Health. 1996;18:301–5.

Moran A et al. Insulin resistance during puberty: results from clamp studies in 357 children. Diabetes. 1999;48:2039–44.

Moran A et al. Association between the insulin resistance of puberty and the insulin-like growth factor-I/growth hormone axis. J Clin Endocrinol Metab. 2002a;87:4817–20.

Moran A et al. Association between insulin resistance of puberty and insulin-like growth factor-I/growth hormone axis. J Clin Endocrinol Metab. 2002b;87:4817–20.

Pijl H et al. Altered neuroregulation of GH secretion in viscerally obese premenopausal women. J Clin Endocrinol Metab. 2001;86:5509–15.

Powell D et al. Insulin inhibits transcription of the human gene for insulin-like growth factor-binding protein-1. J Biol Chem. 1991;266(28):18868–76.

Pulido J, Salazar M. Changes in insulin sensitivity, secretion and glucose effectiveness during the menstrual cycle. Arch Med Res. 1999;30:19–22.

Qu X-D et al. Influence of body mass index and gender on growth hormone (GH) responses to GH-releasing hormone plus arginine and insulin tolerance tests. J Clin Endocrinol Metab. 2004;90:1563–9.

Reynolds R et al. Effect of the glycemic index of carbohydrates on Acne vulgaris. Nutrients. 2010;2:1060–72.

Schaefer O. Pre- and post-natal growth acceleration and increase sugar consumption in Canadian Eskimos. Can Med Assoc J. 1970;103:1059–68.

Schaefer O. When the Eskimo comes to town. Nutr Today. 1971;6:8–16.

Shaw J, White L. Persistent acne in adult women. Arch Dermatol. 2001;137:1252–3.

Singh A et al. Effect of insulin-like growth factor-type I (IGF-I) and insulin on the secretion of sex hormone binding globulin and IGF-I binding protein (IBP-I) by human hepatoma cells. J Endocrinol. 1990;124:R1–3.

Smith R et al. A low glycemic load diet improves symptoms in acne vulgaris patients. A randomized controlled trial. Am J Clin Nutr. 2007a;86:107–15.

Smith R et al. The effect of a high protein, low glycemic load diet versus a conventional, high glycemic load diet on biochemical parameters associated with acne vulgaris. A randomized, investigator-masked, controlled trial. J Am Acad Dermatol. 2007b;57:247–56.

Stathakis V, Kilkenny M, Marks R. Descriptive epidemiology of acne vulgaris in the community. Australas J Dermatol. 1997;38:115–23.

Sydney University: The official website of the glycemic index and GI database. 2010. http://www.glycemicindex.com. Accessed 2010.

Thiboutot D, Lookingbill D. Acne: acute or chronic disease? J Am Acad Dermatol. 1995;32:S2–5.

US Department of Health Education and Welfare. Skin conditions of youths 12–17 years. United States. In: Data from the National Health Survey. Series 11. Number 157. DHEW Publication No. 76–1639. Rockville: DHEW; 1976.

Welsh S, Davis C, Shaw A. USDA's food guide: background and development. United States Department of Agriculture, Human Nutrition Information Service. Publication Number 1514. Hyattsville: MD; 1993.

White C. Acneform eruptions of face: etiologic importance of specific foods. J Am Med Assoc. 1934;103:1277–9.

Williams T et al. Impaired growth hormone response to growth hormone-releasing factor in obesity. A pituitary defect reversed with weight reduction. N Engl J Med. 1984;311:1403–7.

Willis D et al. Modulation by insulin of follicle stimulating and luteinizing hormone action in human granulosa cells of normal and polycystic ovaries. J Clin Endocrinol Metab. 1996;81: 302–9.

Wise F, Sulzberger M. The role of the general practitioner in the modern treatment of acne vulgaris. In: The 1933 year book of dermatology and syphilology. Chicago: The Year Book Publishers; 1933. p. 7–12.

Chapter 11
Essential Fatty Acids and Atopic Dermatitis

Anthony Vincent Rawlings

Core Messages

- Atopic dermatitis is a chronic inflammatory skin condition.
- It is associated with many problems concerning the structure and function of the stratum corneum, especially the skin barrier function.
- Deficiencies of downstream metabolites of linoleic acid are associated with the condition.
- Intervention studies suggest that omega-6 fatty acids, especially γ-linolenic acid, may be useful for mild atopic conditions.
- Orally ingested and topically applied oils containing γ-linoleic acid improve skin barrier functioning.

11.1 Abnormalities in the Stratum Corneum in Atopic Dermatitis

11.1.1 Atopic Dermatitis and Impaired Barrier

Atopic dermatitis (AD) is a chronic inflammatory skin disease associated with cutaneous hyperreactivity to environmental triggers (Cork et al. 2006, 2009). It can be categorized into intrinsic and extrinsic subtypes (Tokura 2010). Extrinsic AD is the most severe phenotype, occurs the most frequently, and shows the highest levels of serum immunoglobulin E (IgE). Conversely, the incidence of intrinsic AD is

A.V. Rawlings (✉)
AVR Consulting Ltd, Northwich/Cheshire CW9 8FH, UK
e-mail: TonyRawlings@aol.com

A. Pappas (ed.), *Nutrition and Skin: Lessons for Anti-Aging, Beauty and Healthy Skin*,
DOI 10.1007/978-1-4419-7967-4_11,

approximately 20%; it tends to have a predominance among women and is a less severe condition. The transitional nature of intrinsic AD has led many groups to conclude that a skin barrier defect plays a role in the development of AD (Cork et al. 2006; Elias et al. 1999; Taieb 1999; Bieber 2008; Callard and Harper 2007). In 80% of subjects with intrinsic AD, IgE levels are subsequently elevated, and patients develop extrinsic AD (Bieber 2008; Illi et al. 2004). However, the remaining 20% of patients never develop a high IgE level and continue with the intrinsic AD phenotype (Bieber 2008). These findings support a nonimmune causative event early in the development of AD, such as a defective skin barrier.

An impaired epidermal barrier allows penetration of potential allergens through the skin, thereby facilitating their interaction with the local antigen-presenting cells and immune effector cells. In very mild, permanently intrinsic AD, the "outside–inside hypothesis" (Elias et al. 1999, 2008) may explain the entire disease process. Conversely, in AD that starts as intrinsic but then switches to extrinsic, both the "outside–inside hypothesis" (Cork et al. 2006) and the "inside–outside hypothesis" (Leung 2000; Elias and Steinhoff 2008) may explain different aspects of the disease process at different points in its development. Thus, barrier function might drive disease activity rather than be an epiphenomenon associated with the immunological processes involved in this disease.

Nevertheless, Kikuchi et al. (2006), in a prospective study on newborn infants with family histories of atopic tendencies, reported that transepidermal water loss (TEWL) values were normal, but they became elevated once AD lesions developed. Conversely, reduced skin hydration was observed, and these authors concluded that barrier impairment in AD is not inherent but represents a phenomenon secondary to eczematous skin changes. Skin surface hydration on the flexor forearms of the infants who developed AD, however, tended to be lower than that of the non-AD infants after 1 month.

The resolution of AD can be variable. It may resolve at an early age or during the teenage years, but it may also persist into adult life. The timing of the resolution of the disease may be related to the maturation of the barrier. At birth, the skin barrier has relatively poor function, but this naturally improves as the child becomes older (Nikolovski et al. 2008). Eventually, the lipid component of the SC barrier is lessened with age and with the seasons (Rogers et al. 1996). External factors also influence barrier function. Environmental agents that can perturb barrier function include soap, detergents, olive oil, excessive use of topical corticosteroids, bacterial infection, inhalant allergens such as house dust mites, and all other topical formulations used to treat AD (Cork et al. 2006, 2009). Exposure to soap and detergents has been recognized as an exacerbating environmental factor in AD for four decades. The detrimental effects arise through damage to the lipid lamellae, washout of innate antimicrobial compounds and natural moisturizing factor (NMF), increased pH, and changes in proteases levels and their activities. Even some emollient creams containing high concentrations of surfactants, such as aqueous cream, have been shown to induce irritant reactions in most children attending a pediatric AD clinic (Cork and Danby 2009). Aqueous cream was designed as a wash-off emollient soap substitute (containing 1% sodium lauryl sulfate, or SLS), not as a leave-on emollient, and the damaging effects it has on the skin barrier are an illustration of the hazards of using a topical product incorrectly (Cork and Danby 2009). Aqueous cream even induced damage to the normal skin barrier by thinning the stratum corneum (SC) when applied twice per day for 4 weeks (Tsang and Guy 2010).

11.1.2 Atopic Dermatitis and Aberrations in Stratum Corneum Structure and Function

Elevated TEWL and reduced skin hydration are signs of impaired SC function and are associated with AD (Tagami et al. 2006). Barrier function is related to the total architecture of the SC. The original "bricks and mortar" model has been refined over the years and is now recognized as a continuous polyproteinaceous structure of varying thickness—the bricks are tightly interconnected by corneodesmosomes in all layers of the SC but are actually more like NMF-containing keratin sponges because they hydrate extensively—interspersed between a continuous highly ordered lamellar and largely ortho-rhombically packed lipid phase (Rawlings et al. 1994; Rawlings 2003, 2010; Rawlings and Matts 2005). Clearly, the SC thickness and corneocyte size predominantly control the tortuosity of the SC, although this is also influenced by the swelling of the SC due to the presence of NMF, whereas the corneocyte covalently bound and free intercellular lipids consisting of predominantly ceramides, fatty acids, and sterols provide the waterproofing of the SC. Naturally, all these mechanisms are influenced detrimentally in AD.

Some of the earliest reported differences in SC composition between healthy and atopic subjects were in SC lipid levels. Melnik et al. (1988) first reported the reductions in the levels of ceramides in noneczematous dry skin of atopic individuals in 1988. Yamamoto et al. (1991) soon after observed the reductions in the levels of CER EOS and an increase in its oleate fraction. At about the same time, Imokawa et al. (1991) also reported decreased levels of ceramides (CER EOS and CER NP) in both involved and uninvolved skin of atopic persons. Similar results were reported by Matsumoto et al. (1999), with a 50% reduction in the levels of CER EOS in atopic skin compared with that of healthy subjects. Di Nardo et al. (1998) also found reduced levels of CER EOS and CER NP, whereas cholesterol levels were significantly higher in subjects with AD. Macheleidt et al. (2002) performed lipogenesis studies on biopsy specimens from healthy skin and lesional skin and demonstrated reduced synthesis of all ceramides, except CER NS, and particularly its short chain *N*-acyl fatty acid variant in specimens from the atopic lesional skin. Generally, there was an increase in the ratio of the sphingosine-containing ceramides to the phytosphingosine-containing ceramides, which appears to be characteristic of hyperproliferative skin disorders (Rawlings 2003). Macheleidt et al. (2002) also studied the deficiency of corneocyte protein-bound omega-hydroxyceramides in atopic dermatitis, which are present at 46–53 wt% of total protein-bound lipids in healthy skin. Their levels were reduced to 23–28% and 10–25% in nonlesional and lesional areas, respectively. These changes resulted in an increase in the corneocyte-bound omega-hydroxy fatty acid fraction and smaller increases in the fatty acid fraction. Furthermore, the lipids present had a "hydrocarbon chain length deficiency" with reduced presence of very-long-chain fatty acids with more than 24 carbon atoms. Bleck et al. (1999) studied the noninvolved skin of atopic eczema (NEAE) and found that all ceramide species were decreased, especially CER EOS and CER NP, whereas the CER EOH levels were increased. After slightly modifying the chromatographic solvent conditions, two species could be observed in the CER AS position; containing C_{16-18} and $C_{22, 24, 26}$ α-hydroxy fatty acids. This seemed to be specific for AD, as only a single peak was found in samples from senile xerosis, seborrheic eczema, and psoriasis tissues. Nevertheless, they did not observe increases in the cholesterol fraction. Farwanah et al. (2005) could not confirm these results. More recently, Ishikawa et al. (2010) found that the levels of CER EOS, EOH, EOP, NP, and NH to be lower in atopic subjects. They also found that the larger ceramide species of >50 carbon atoms of CER NS, NDS, NH, AS, and AH had lower expression, whereas the smaller species of CER NS, NDS, and AS tended to be expressed at higher levels. Di Nardo et al. (1998) previously found a negative correlation with TEWL and the quantity of CER NP and a positive correlation with AH. In that study, a strong correlation with impaired barrier function was observed with the level of CER NS and EOP. Positive correlations with improved barrier function were observed with CER NDS, NH, AH, AP, EOS, EOH, and EOP. Some of the epidermally produced lipids are also reported not to be transported to their correct intercorneoctytic location. A disturbance of extrusion of lamellar bodies and fusion of intercellular lipids has been reported by Fartasch et al. (1992) in the dry noneczematous skin of AD patients.

A "hydrocarbon chain length deficiency" appears to occur in the ceramide and free fatty acid (FFA) fraction of SC lipids in patients with AD. Schafer and Kragballe (1991) showed that the content of long-chain fatty acids (LCFAs) versus short-chain fatty acids (SCFAs) was decreased in epidermal lipids of lesional versus lesion-free AD subjects. However, this hydrocarbon chain length deficiency appears to be an effect of general hyperproliferative diseases as Nicollier et al. (1986) reported a decreased presence of LCFAs in hyperkeratotic corneum. In other, milder forms of barrier disruption, Fulmer and Kramer (1986) found decreased levels of C22–28 SC fatty acids in SLS-induced dry skin. Brod et al. (1988) also could also not detect any long-chain species in the free and esterified fatty acids of dry skin. Nevertheless, the importance of these long-chain species has only recently become apparent from the work of Bouwstra and colleagues.

Ceramides are responsible for the lamellar packing of SC lipids, and their inter-bilayer spacing is very much dictated by the presence of CER EOS, where a long periodicity phase (LPP) is present and with sufficiently attached linoleate, a fluid phase. In the absence of the LPP, a short periodicity phase (SPP) predominates (Bouwstra et al. 1998; Bouwstra and Ponec 2006). These lamellar spacings are still present with cholesterol, yet the lateral packing state is predominantly a hexagonal one. A much tighter orthorhombic packing state, responsible for a lower TEWL, is induced in the presence of long-chain but not short-chain fatty acids. Although the hexagonal lateral packing state of the SC lipids is known in subjects with AD (Pilgram et al. 2001), the presence of the LPP has not yet been defined fully.

However, most recently, Janssens et al. (2011) have demonstrated heterogeneity in the LLP of Sc in non-involved atopic subjects with the suggestion of an increased SPP at the expense of LPP.

The biochemical reason for the reduced VLCFA species in AD is not yet fully understood, although Saaf et al. (2008) observed down-regulation of lipid biosynthesis genes in AD. Not only did they find reductions in the levels of delta-5 (FADS1) and delta-6 (FADS2) desaturases (enzymes involved in essential fatty acid metabolism see later), but there was also decreased levels of the elongase enzymes—e.g., elongase of very-long-chain fatty acid-5 (ELOV5). It is likely that other elongases are not transcribed properly in AD and lead to a deficiency in the synthesis of LCFAs. Indeed, barrier defects are known in mice carrying the Stargardt disease 3 mutation, which has a deficiency in elongase of very-long-chain-fatty acid-4 (ELOVL4) which leads to an absence of acylceramides as well as LCFAs. In this disorder, C26 acyl length species accumulate (McMahon et al. 2007; Cameron et al. 2007; Vasireddy et al. 2007).

Ceramide deficiencies may also occur via degradation of ceramides within the SC. At least three enzyme activities can lead to reduced ceramide levels in the SC. Ceramidase, glucosylceramide deacylase, and sphingomyelin deacylase are known to hydrolyze the corresponding sphingolipid precursors at their *N*-acyl bonds to yield sphingosine, glucosyl sphingosine, and sphingosylphosphorylcholine together with the corresponding FFAs. Hara et al. (2000) and Arikawa et al. (2002) have reported that acid ceramidase is significantly down-regulated in atopic skin, whereas Higuchi et al. (2000) and Hara et al. (2000) found increased levels of sphingomylein deacylase. Similarly, Okamato et al. (2003) also found increased levels of sphingosylphosporylcholine in atopics. Ishibashi et al. (2003) found increased glucosyl ceramide deacylase. Imokawa (2009) has proposed that these two enzyme activities reside in the same enzyme: sphingomyelin glucosylceramide deacylase.

These differences in lipid levels has led to the assumption that patients with AD have an inherent impairment of SC barrier function that facilitates the penetration of allergens (Cork et al. 2009). However, changes in other SC components can equally be involved. There are now many reports of filaggrin mutations being associated with extrinsic AD (Brown and Irvine 2008). Reduced levels of filaggrin lead to reduced NMF levels, which probably accounts for the reduced skin hydration characteristics of AD skin. In its broadest sense, filaggrin is also required for maturation of the SC, and thereby optimal SC barrier function, and such elevated TEWL levels can be found in subjects with loss of function mutations in the filaggrin gene as reported by Kezic et al. (2008). Importantly, by studying the flaky-tail mouse model, which lacks processed filaggrin, it was observed that a paracellular barrier abnormality occurred with reduced inflammatory thresholds to topical haptens, further highlighting the importance of filaggrin for the formation of a fully functioning SC (Scharschmidt et al. 2009).

The differences in SC lipids in subjects with filaggrin mutations and SC lipids are complex compared with healthy subjects. For instance, Jungersted et al. (2010) examined SC lipids with filaggrin mutations and atopic eczema and found significantly lower levels of ceramide 4 (CER EOH) and higher levels of ceramide 7 (CER AP) in the AD group with filaggrin mutations. Equally, the AD group had lower CER EOS.

Changes in SC thickness can occur in subjects with AD. At this point, we need to consider the subtle dryness of the skin surrounding lesions of AD differently. It has been called atopic xerosis (AX). AX is more susceptible to the development of AD lesions than clinically normal skin. Tagami et al. (2006) reported that the number of SC cell layers in AX is significantly greater than in normal skin, whereas White et al. (1987) reported that the number of SC cell layers is significantly lower in patients with chronic eczematous AD conditions. These differences can only come from changes in the balance between desquamation and epidermal proliferation.

Desquamation occurs via corneodesmolysis—i.e., proteolysis of the SC corneodesmosomes (Rawlings et al. 1994; Rawlings 2003, 2010; Rawlings and Matts 2005; Harding et al. 2000). Corneodesmosomes are specialized desmosomes holding the corneocytes together via transmembrane cadherin proteins (desmoglein 1 and desmocollin 1) together with corneodesmosin and desmosealin. Differences in the structure of the corneodesmosomes in subjects with AD has been reported by Pilgram et al. (2001) However, their proteolysis occurs through the concerted action of a variety of proteases, not least of which are the kallikreins and especially the SC chymotrypsin-like and trypsin-like kallikrein. Their complexity has been further unravelled by Komatsu et al. (2006), who confirmed their presence and identified particular kallikreins (KLKs) in SC extracts immunologically (KLK5, KLK6, KLK7, KLK8, KlK10, KLK11, KLK13, KLK14). They are all believed in some way to play a role in corneodesmolysis, and in this respect the SC trypsin-like activities are important. Redoules et al. (1999) observed reduced activity of SC trypsin-like kallikreins in dry noneczematous AD skin (i.e., AX lesions). As this class of enzyme is essential for the degradation of desmoglein 1, its reduced activity may explain the increased number of SC cell layers reported by Tagami et al. (2006) Equally, Tarroux et al. (2002) demonstrated that its activity increases as the lesions are resolved. Conversely, Komatsu et al. (2007) found increased mass levels of SC kallikreins together with plasmin and furin in mild lichenified AD lesions, which may be attributed to the decreased number of cell layers in AD lesions reported by White et al. (1987) Equally, Hansson et al. (2002) reported increased SCCE (KLK7) immunostaining in skin biopsies taken from AD subjects with chronic eczematous lesions on the flexural sides of the lower arms. Moreover, Vasilopoulos et al. (2004) identified a 4-bp AACC insertion in the 3′ untranslated region of the KLK7 gene, especially in subjects who did not have elevated levels of IgE (intrinsic AD). It was proposed that this insertion could increase the half-life of KLK7 mRNA, leading to increased levels of KLK7 in affected individuals. Most recently, elevated extractable serine protease activity was measured in AD skin in the order: SC tryptase-like enzyme, plasmin, trypsin-like kallikreins, urokinase, chymotrypsin-like kallikreins, and leukocyte elastase activity (Voegeli et al. 2009, 2011). A significantly thinner was SC was reported in the presence of these extra enzyme activities (Voegeli et al. 2009). Saaf et al. (2008) recently reported increased gene expression of KLK7/8 and corneodesmosin (Cdsn) together with decreased expression of Dsg2.

Decreased serine protease inhibitors may account for the above activities. The most compelling evidence for the role of excess serine protease activity due to reduced levels of SC-derived serine protease inhibitors in the pathogenesis of AD in humans comes from Netherton syndrome (NS) (Descargues et al. 2006). NS includes AD as one of its manifestations. Mutations in the serine protease inhibitor Kazal-type 5 (SPINK5) gene, which encodes the lymphoepithelial Kazal-type 5 serine protease inhibitor (LEKTI), have been linked to NS (Kato et al. 2003). Mutations in the SPINK5 gene have also been associated with AD. Reduced levels of LEKTI leads to a thinner SC because of uncontrolled serine protease degradation of the corneodesmosomes. In the AD animal model reported by Man et al. (2008), serine protease activity occurred throughout the entire SC. Additionally, corneocytes appeared to detach prematurely between the SC and the underlying nucleated cell layers.

In AX, the presence of immature corneocytes may also contribute to decreased skin barrier function via their detrimental effect on the tortuosity of the stratum corneum. Immature and mature corneocyte envelopes (CEs) can be differentiated by their binding of tetramethylrhodamine isothiocyanate (TRITC), with the rigid envelopes staining to a greater extent (Harding et al. 2003) or based on their hydrophobicity (staining with Nile red) and antigenicity (to anti-involucrin) (Hirao et al. 2001). The maturity index of the corneocytes is important, as smaller immature corneocytes are associated with barrier-compromised conditions. Hirao et al. (2003) reported on four types of CE with and without the presence of parakeratosis in AD. The immaturity can also be assessed by atopic force microscopy (Kashibuchi et al. 2002). Differences in the presence of these CEs reflect aberrant keratinocyte proliferation and differentiation together with abnormal expression of transglutaminases.

Thus, all aspects of SC composition and architecture can be observed to change in subjects with AD. Changes in SC thickness and corneocyte size leads to reduced SC tortuosity, which can be further reduced by reductions in the levels of NMF. Selected reductions and increases in SC lipids lead to a weak hexagonally packed lipid barrier. The role of the LPP needs to be further defined, but changes are expected in lesional skin, as CER EOS linoleate levels are known to be decreased in subjects with AD.

11.2 Importance of Omega-6 EFA in Atopic Dermatitis

Burr and Burr were the first to define essential fatty acid deficiency (EFAD) (Burr and Burr 1929, 1930). Hansen et al. (1933, 1958), and Hansen and Wiese (1954) much later speculated that atopic eczema may be an EFAD as many of the skin symptoms were similar. Also, Brown and Hansen (1937) detected decreased levels of arachidonic acid (AA) in the serum of children with AD. Equally, examining blood samples, Manku et al. (1982) did not find a linoleic acid deficiency but did find a deficiency of their downstream metabolites: γ-linolenic acid (GLA), dihomo-γ-linolenic (DGLA), and AA. This suggested that there might be a deficiency in the levels of the desaturases needed to perform these biochemical conversions in atopic subjects. Manku et al. (1982) and Horrobin (1989, 2000) defined the hypothesis that the fatty acid profiles in subjects with AD reflected lower activities of the desaturase enzymes. The liver can biotransform LA to GLA, but its rate of transformation can be reduced (e.g., with aging or in subjects under psychological stress; Fan and Chapkin 1998). Human skin is not able to biosynthesize GLA from its precursor acid LA because of a lack of the delta-6 desaturase (Saaf et al. 2008; Ziboh et al. 2002) and is therefore especially sensitive to changes in the blood levels of GLA. The delta-5 and delta-6 desaturases are considered the rate-limiting enzymes in the formation of long-chain polyunsaturated fatty acids (PUFAs) (Innis 2003; Nakamura and Nara 2004; Sprecher et al. 1995). The genes have been identified as FADS1 and FADS2 (Marquardt et al. 2000) Linoleic acid is converted by delta-6 desaturation to GLA, which is followed by elongation to DGLA. The delta-5 desaturase converts DGLA to AA. Park et al. (2009) reported an alternate pathway via elongation of linoleic acid to eicosadienoic acid followed by a delta-8 desaturation to DGLA.

Galli et al. (1994) have also shown in infants that the aberrant EFA profiles preceded the development of AD. GLA is the precursor to prostaglandin E_1 (PGE_1), which is antiinflammatory; and when humans consume a diet high in GLA, its metabolic products PGE_1 and 15-hydroxy-eicosatrienoic acid levels accumulate in skin (Ziboh et al. 2002; Ziboh and Chapkin 1987; Ziboh and Miller 1990; Ziboh 1994).

Saaf et al. (2008) recently demonstrated reduced expression levels of the genes for FAD1 FAD2, supporting the hypothesis that the lower levels of PUFA metabolites in AD are due to impaired synthesis. Schaeffer et al. (2006) analyzed the single nucleotide polymorphisms (SNPs) of the FADS1 and FADS2 gene cluster. Carriers of the minor alleles of 11 SNPs showed enhanced levels of LA, EDA and DGLA and decreased levels of GLA and AA. Subjects carrying minor alleles of several SNPs FADS1 and FADS2 had lower levels of atopic eczema, but it was not related to IgE levels.

γ-Linolenic acid (18:3n-6) is present in epidermal glucoceramides (Chung et al. 2002). However, only very small quantities of GLA are present in SC ceramides (Colarow 1990). Much higher quantities are found in epidermal glucoceramides, suggesting that further metabolism of GLA in epidermal ceramides together with other lipids is essential for epidermal differentiation. To understand the relation between EFAs and AD, animal models of EFAD can be utilized. Studies on EFAD animals have demonstrated that LA is essential for repairing the impaired barrier function in such conditions (Houtsmuller and van der Beek 1981). However, GLA induces barrier repair more quickly than LA by at least a day, again implying that further metabolites of GLA are important for the epidermal differentiation process (Hartop and Prottey 1976). These may be DGLA (20:3n-6) or 15-hydroxy-eicosatrienoic acid (15-HETrE). The 1-series of prostaglandins are important antiinflammatory lipids that are also derived from GLA, but they do not directly induce barrier repair in EFAD models (Prottey 1977).

Borage oil-derived GLA, when supplied in the sn-2 position of the triglyceride oil, is more effective than evening primrose oil-derived GLA, which is present in the sn-3 position of the triglyceride. It was utilized in reversing epidermal hyperproliferation and increasing ceramide synthesis in guinea pigs induced into an EFAD state using a hydrogenated coconut diet (HCO) for 8 weeks. In that study, guinea pigs were fed HCO (14 g/kg) to induce the deficiency state (Chung et al. 2002). To reverse the EFAD, the animals were fed for 10 weeks with primrose oil (PO), borage oil (BO), or equal quantities of BO+PO (BS) at doses of 60 g/kg. The combination treatment was designed to provide a level of GLA similar to that of PO but in a different position on the triglyceride backbone. All three GLA-containing diets suppressed epidermal thickening induced by the HCO diet, although the BO diet was most effective. Upon examination of thymidine incorporation into epidermal DNA, the GLA diets were effective but the BO diet was best. GLA was not detected in epidermal lipid fractions, but DGLA was. DGLA showed greater incorporation into epidermal phospholipids and ceramides in the order BO>BS>PO. The borage oil diets also showed greater incorporation of linoleic acid into epidermal ceramides. These results suggest that the absolute level of GLA in oils determines the accumulation of DGLA in epidermal phospholipids and ceramides and that the presence of GLA in the sn-2 position of the triglyceride oil mediates this whereas the presence of LA in the sn-1 position of the triglyceride facilitates the accumulation of linoleic acid in epidermal ceramides. The content of other antiinflammatory and antiproliferative metabolites of these lipids (13-HODE, HETrE) reflect their incorporation into the epidermal lipid fractions. Finally, stimulation of epidermal ceramide synthesis was greater for the borage oil diet alone, in the order BO>BS>PO. In these studies, a sunflower oil (SO) diet alone was inferior to the PO diet and was significantly inferior to the BO diets. On the SO diet, only 0.7 g of DGLA/100 g of total epidermal fatty acids was found in the skin. In comparison, the PO diet delivered 2.9 g, the BS diet delivered 11.7 g, and the BO diet delivered 15.2 g. These gave DGLA/LA ratios of 0.9%, 5.0%, 17.4%, and 27.5%, respectively. Thus, because of its more efficient bioavailability, borage oil was more effective than evening primrose oil in these studies. More importantly, however, as skin cannot make GLA from LA because it lacks the necessary desaturase enzyme to perform this metabolic step, it is highly dependent on the blood for its supply of GLA and subsequently its further metabolites. Although the precise mechanism of action of GLA has not been determined in these studies, it is believed to be a direct effect of GLA itself on the keratinocyte differentiation process or due to the effects of its downstream metabolites (DGLA or 15-HETrE). Nevertheless, its effects are probably due to improving the epidermal differentiation process by influencing relevant transcription factors—e.g., activator protein-1 (De Pascale et al. 2006) and/or peroxisome proliferator activated receptors (PPARs) (Jiang et al. 2000).

11.3 Effect of Omega-6 EFAs on Atopic Dermatitis

Provision of appropriate essential fatty acids (EFAs) should be expected to correct the skin defects attributable to the EFA deficiency in AD. If in the presence of atopic eczema there is a reduced rate of conversion of linoleic to GLA, it makes sense to provide oils enriched with GLA. However, these are not drugs but nutrients, and the doses usually applied to humans are low compared to the dosages used in animal studies. As a result, it is likely that they provide benefit only for the mild AD disorders, possibly intrinsic AD, and not the severe extrinsic disorders if only used orally.

Applying the ingredients topically is likely to deliver more effect. This section reviews what studies have been conducted in subjects with AD and other milder forms of barrier dysfunction where they might also be appropriate.

Several oral intervention studies have been conducted to assess the effects of EFAs on the alleviation of AD. However, Van Gool et al. (2004) recently performed a meta-analysis of placebo-controlled EFA trials and came to the conclusion that the effects of EFAs were negligible; that is, they did not exert a large effect on the clinical condition. Equally, Foster et al. (2010) performed a similar analysis of the effects of borage oil and also concluded that a major clinical effect is unlikely to be seen, although it may be useful in some patients with less severe AD and might be used as maintenance treatment to prevent flare-ups in mild disease states. In this respect, EFAs have been used as preventative therapy in infants as EFA abnormalities precede the development of AD. Van Gool et al. (2003) fed formula to babies at high risk for AD that contained borage oil or sunflower seed oil. Although AD was not prevented, there was a trend for less symptoms in the borage oil group.

Furthering these positive results, Linnamaa et al. (2010) compared the effects of black currant seed oil to olive oil supplements to neonates from mothers at high risk for atopy. The supplements were continued until cessation of breast-feeding and were followed by direct supplementation to the infants until 2 years of age. There was a significantly lower prevalence of AD in the black currant-supplemented group, with lower SCORAD scores at 12 months of age. Although this significance was lost at 24 months, there was still a trend toward a reduced incidence of disease. These effects are similar to those of Johansson et al. (1999), who compared the effects of alpine currant seed oil with rapeseed oil and found that the intensity of skin itching and other dermatitis symptoms were reduced in children up to 4.5 years. It is possible that these positive effects are lost as a less healthy diet begins to be adopted or there is poor skin care leading to skin barrier problems (e.g., use of aqueous cream with SLS as a primary emulsifier, which is known to have detrimental effects on skin barrier function; Tsang and Guy 2010). Kitz et al. (2006) came to the same conclusion in infants suffering from AD. GLA supplementation seemed to reduce the total IgE concentration during the first year of life. Callaway et al. (2005) conducted a randomized single-blind crossover study on AD subjects in 2005 using hempseed oil compared with olive oil. Skin dryness and skin itching were reduced, and there was a trend toward improved skin barrier function. Hempseed oil, however, also contains the omega-3 fatty acid α-linolenic acid. Senapati et al. (2008), however, also demonstrated that EPO was effective compared to SSO in Indian subjects.

One of the problems with the relative low efficacy of EFA supplements in the more severe AD conditions may be related to the poor bioavailability of the EFAs. Emulsification of the oils may aid delivery. In fact, Puch et al. (2008), studying normal, healthy subjects, demonstrated pharmacokinetic changes after emulsification compared to the free oil. A faster T_{max} and higher area under the curve were observed.

The enhanced oral bioavailability allowed the GLA to have a greater effect on epidermal differentiation. Indeed, relative to placebo, skin barrier function was significantly improved after consumption of the product. The authors could not rule out the effects of smaller quantities of catechins and vitamin E, but the bulk of the benefit was probably derived from the borage oil. These findings were further exemplified by comparison to the results of others. The study of Broche and Platt (2000) showed improved barrier function after supplementation with 320 and 740 mg of GLA in elderly subjects; and Muggli et al. (2005) used 300 mgsin a younger population with similar results. Respectively, the relative improvement in TEWL to baseline was 10.8% (7.65 vs. 6.82 $g/m^2/h$ at 8 weeks; $P<0.05$); and relative to placebo it was 7.7% and 9.1% (9.1 vs. 8.4 $g/m^2/h$ and 8.7 vs. 7.9 $g/m^2/h$ at 12 weeks, $P<0.05$). In the Puch et al. study the effect on TEWL occurred more quickly (6 weeks), and the percentage improvement in barrier function relative to placebo was greater (13.25% in the total group and 15.0% in the subjects with body mass index of <25) presumably due to the improved bioavailability of at least the GLA in the product.

Improvements in SC NMF levels were also demonstrated using in vivo confocal raman spectroscopy by Krahn-Bertil et al. (2009) and Rawlings et al. (2011). A threefold higher percentage improvement was observed for the borage oil, green tea, and vitamin E emulsion than in the control group (control group +4.2%; test group +11.44%). Further studies on subjects with AD are advocated with these new findings.

Surprisingly, there are few reported studies of the effects of EFAs applied topically to AD patients. Gehring et al. (1999) established the effect on barrier function in AD using topical evening primrose oil in an amphiphilic and a stable water-in-oil emulsion. The studies were vehicle-controlled in two populations of 20 atopic subjects, and barrier function was assessed in terms of TEWL and SC hydration after a 4-week treatment period and a 1-week treatment-free period. Evening primrose oil proved to have a stabilizing effect on the SC barrier, but this was apparent only with the water-in-oil emulsion, not the amphiphilic emulsion. The choice of vehicle is therefore an extremely important factor in the efficacy of topically applied evening primrose oil. Most recently, topical borage oil coated onto undershirts has been shown by Kanehara et al. (2007a, b) to alleviate the symptoms of AD in children.

Other studies have demonstrated the efficacy of EFAs in barrier-compromised infant skin. Sunflower seed oil significantly accelerated skin barrier recovery within 1 h; and the effect was sustained 5 h after application. In contrast, the other vegetable oils tested (mustard, olive, and soybean oils) significantly delayed recovery of barrier function compared with control skin or Aquaphor-treated skin (Darmstadt et al. 2002).

In normal, healthy adults, topical borage and sunflower seed oils have been shown to be beneficial in winter when the levels of CER EOS linoleate are reduced (Conti et al. 1996). After treatment, the levels of CER EOS linoleate were normalized. Presumably in these studies, the linoleate is delivered to the epidermis and used as a precursor in the CER EOS biosynthetic pathway. Following the metabolism of epidermal linoleate in vivo, Wertz and Downing (1990) showed that most labeled linoleate was initially associated with phospholipids, triacylglycerols, and FFAs. After, 3–7 days, the label transferred to acylglucosylceramide and then to CER EOS. Thus, linoleate transfers from phospholipid to FFAs to acylglucosylceramide and finally to the ceramides. In vitro studies have shown that glucosylation of CER EOS occurs before omega oxidation of the *N*-acylated fatty acid and presumably esterification with linoleic acid (Uchida and Holleran 2008). Whether this sequence of events was operating in the studies of Conti et al. (1996) is unknown. However, the hydrolyzed EFAs from the oils may also be acting as PPAR agonists to induce epidermal differentiation, as discussed earlier. As such, increased levels of all SC components are found, including filaggrin and thus NMF (Harding and Rawlings 2006).

Take Home Messages

- Atopic dermatitis (AD) is a complex disease that is associated with many aberrations of the stratum corneum (SC), including lipid abnormalities, reductions in filaggrin levels and thus NMF levels, excessive or lessened amounts of proteases and reduced levels of their intrinsic inhibitors, a thinner and/or thicker SC together with the presence of less-mature corneocytes.
- Altered gene expression is known; but decreases in the levels of the delta-5 and delta-6 lipid desaturase enzymes are particularly important. As a result, downstream metabolites of linoleic acid are missing in skin.
- Ingestion of essential fatty acids, particularly γ-linolenic acid, is known to improve the epidermal differentiation associated with essential fatty acid deficiency. AD can be considered an essential fatty acid insufficiency state.
- Use of omega-6 rich oils in meta-analyses have been reported to be of little use for the extreme forms of AD. However, they are proposed for the milder (intrinsic) forms of AD and have in particular been shown to prevent or reduce the occurrence of the disease. Much more research is needed, but enhanced oral delivery using emulsification methods may aid its efficacy.
- Clearly, topically applied omega-6-containing oils can alleviate the disease state and possibly correct some of the deficiencies known to occur in AD, which include a deficiency of CER EOS linoleate and filaggrin together with natural moisturizing factor.

References

Arikawa J, Ishibashi M, Kawashima M, et al. Decreased levels of sphingosine, a natural antimicrobial agent, may be associated with vulnerability of the stratum corneum from patients with atopic dermatitis to colonization by Staphylococcus aureus. J Invest Dermatol. 2002;119:433–9.

Bieber T. Atopic dermatitis. N Engl J Med. 2008;358:1483–94.

Bleck O, Abeck D, Ring J, et al. Two ceramide subfractions detectable in Cer(AS) position by HPTLC in skin surface lipids of non-lesional skin of atopic eczema. J Invest Dermatol. 1999;113:894–900.

Bouwstra JA, Ponec M. The skin barrier in healthy and diseased state. Biochim Biophys Acta. 2006;1758:2080–95.

Bouwstra JA, Gooris GS, Dubbelaar FE, et al. Role of ceramide 1 in the molecular organization of the stratum corneum lipids. J Lipid Res. 1998;39:186–96.

Brod J, Traitler H, Studer A, et al. Evolution of lipid composition in skin treated with blackcurrant seed oil. Int J Cosmet Sci. 1988;10:149–59.

Brosche T, Platt D. Effect of borage oil consumption on fatty acid metabolism, transepidermal water loss and skin parameters in elderly people. Arch Gerontol Geriatr. 2000;30:139–50.

Brown WR, Hansen AE. Atachidonic acid and linoleic acid of the serum in normal and eczematous human subjects. Proc Soc Exp Biol Med. 1937;36:113–6.

Brown SJ, Irvine AD. Atopic eczema and the filaggrin story. Semin Cutan Med Surg. 2008;27:128–37.

Burr GO, Burr MM. A new deficiency disease produced by the rigid exclusion of fat from the diet. J Biol Chem. 1929;82:345–67.

Burr GO, Burr MM. On the nature and role of the fatty acids essential in nutrition. J Biol Chem. 1930;86:587–621.

Callard RE, Harper JI. The skin barrier, atopic dermatitis and allergy: a role for langerhans cells? Trends Immunol. 2007;28:294–8.

Callaway J, Schwab U, Harvima I, et al. Efficacy of dietary hempseed oil in patients with atopic dermatitis. J Dermatolog Treat. 2005;16:87–94.

Cameron DJ, Tong Z, Yang Z, et al. Essential role of Elovl4 in very long chain fatty acid synthesis, skin permeability barrier function, and neonatal survival. Int J Biol Sci. 2007;3:111–9.

Chung S, Kong S, Seong K, et al. Gamma-linolenic acid in borage oil reverses epidermal hyperproliferation in guinea pigs. J Nutr. 2002;132:3090–7.

Colarow L. Quantification of ceramides with essential fatty acids moieties in the human skin surface and blood plasma lipids. J Planar Chromatogr. 1990;3:126–32.

Conti A, Rogers J, Verdejo P, et al. Seasonal influences on stratum corneum ceramide 1 fatty acids and the influence of topical essential fatty acids. Int J Cosmet Sci. 1996;18:1–12.

Cork MJ, Danby S. Skin barrier breakdown: a renaissance in emollient therapy. Br J Nurs. 2009;18:872. 4, 6–7.

Cork MJ, Robinson DA, Vasilopoulos Y, et al. New perspectives on epidermal barrier dysfunction in atopic dermatitis: gene environment interactions. J Allergy Clin Immunol. 2006;118:3–21.

Cork MJ, Danby SG, Vasilopoulos Y, et al. Epidermal barrier dysfunction in atopic dermatitis. J Invest Dermatol. 2009;129:1892–908.

Darmstadt GL, Mao-Qiang M, Chi E, et al. Impact of topical oils on the skin barrier: possible implications for neonatal health in developing countries. Acta Paediatr. 2002;91:546–54.

De Pascale MC, Bassi AM, Patrone V, et al. Increased expression of transglutaminase-1 and PPARgamma after vitamin E treatment in human keratinocytes. Arch Biochem Biophys. 2006;447:97–106.

Descargues P, Deraison C, Prost C, et al. Corneodesmosomal cadherins are preferential targets of stratum corneum trypsin- and chymotrypsin-like hyperactivity in Netherton syndrome. J Invest Dermatol. 2006;126:1622–32.

Di Nardo A, Wertz P, Giannetti A, et al. Ceramide and cholesterol composition of the skin of patients with atopic dermatitis. Acta Derm Venereol. 1998;78:27–30.

Elias PM, Steinhoff M. "Outside-to-inside" (and now back to "outside") pathogenic mechanisms in atopic dermatitis. J Invest Dermatol. 2008;128:1067–70.

Elias PM, Wood LC, Feingold KR. Epidermal pathogenesis of inflammatory dermatoses. Am J Contact Dermat. 1999;10:119–26.

Elias PM, Hatano Y, Williams ML. Basis for the barrier abnormality in atopic dermatitis: outside-inside-outside pathogenic mechanisms. J Allergy Clin Immunol. 2008;121:1337–43.

Fan YY, Chapkin RS. Importance of dietary gamma-linolenic acid in human health and nutrition. J Nutr. 1998;128:1411–4.

Fartasch M, Bassukas ID, Diepgen TL. Disturbed extruding mechanism of lamellar bodies in dry non-eczematous skin of atopics. Br J Dermatol. 1992;127:221–7.

Farwanah H, Raith K, Neubert RH, et al. Ceramide profiles of the uninvolved skin in atopic dermatitis and psoriasis are comparable to those of healthy skin. Arch Dermatol Res. 2005;296:514–21.

Foster RH, Hardy G, Alany RG. Borage oil in the treatment of atopic dermatitis. Nutrition. 2010;26:708–18.

Fulmer AW, Kramer GJ. Stratum corneum lipid abnormalities in surfactant-induced dry scaly skin. J Invest Dermatol. 1986;86:598–602.

Galli E, Picardo M, Chini L, et al. Analysis of polyunsaturated fatty acids in newborn sera: a screening tool for atopic disease? Br J Dermatol. 1994;130:752–6.

Gehring W, Bopp R, Rippke F, et al. Effect of topically applied evening primrose oil on epidermal barrier function in atopic dermatitis as a function of vehicle. Arzneimittelforschung. 1999;49:635–42.

Hansen AE. Study of iodine number of serum fatty acids in infantile eczema. Proc Soc Exp Biol Med. 1933;30:1198–9.

Hansen AE, Wiese HF. Essential fatty acids and human nutrition II. Serum level for unsaturated fatty acids in poorly nourished infants and children. J Nutr. 1954;52:367–74.

Hansen AE, Haggard ME, Boelesche AN, et al. Essensial fatty acids in infant nutrition III. Clinical manifestations of linoleic acid deficiency. J Nutr. 1958;66:565–76.

Hansson L, Backman A, Ny A, et al. Epidermal overexpression of stratum corneum chymotryptic enzyme in mice: a model for chronic itchy dermatitis. J Invest Dermatol. 2002;118:444–9.

Hara J, Higuchi K, Okamoto R, et al. High-expression of sphingomyelin deacylase is an important determinant of ceramide deficiency leading to barrier disruption in atopic dermatitis. J Invest Dermatol. 2000;115:406–13.

Harding CR, Rawlings AV. Effects of natural moisturizing factor and lactic acid isomers on skin function. In: Loden M, Maibach HI, editors. Dry skin and moisturizers. 2nd ed. 2006. p. 187–200. Chapter 18.

Harding CR, Watkinson A, Rawlings AV, et al. Dry skin, moisturization and corneodesmolysis. Int J Cosmet Sci. 2000;22:21–52.

Harding CR, Long S, Richardson J, et al. The cornified cell envelope: an important marker of stratum corneum maturation in healthy and dry skin. Int J Cosmet Sci. 2003;25:157–67.

Hartop PJ, Prottey C. Changes in transepidermal water loss and the composition of epidermal lecithin after applications of pure fatty acid triglycerides to skin of essential fatty acid-deficient rats. Br J Dermatol. 1976;95:255–64.

Higuchi K, Hara J, Okamoto R, et al. The skin of atopic dermatitis patients contains a novel enzyme, glucosylceramide sphingomyelin deacylase, which cleaves the N-acyl linkage of sphingomyelin and glucosylceramide. Biochem J. 2000;350(Pt 3):747–56.

Hirao T, Denda M, Takahashi M. Identification of immature cornified envelopes in the barrier-impaired epidermis by characterization of their hydrophobicity and antigenicities of the components. Exp Dermatol. 2001;10:35–44.

Hirao T, Terui T, Takeuchi I, et al. Ratio of immature cornified envelopes does not correlate with parakeratosis in inflammatory skin disorders. Exp Dermatol. 2003;12:591–601.

Horrobin DF. Essential fatty acids in clinical dermatology. J Am Acad Dermatol. 1989;20:1045–53.

Horrobin DF. Essential fatty acid metabolism and its modification in atopic eczema. Am J Clin Nutr. 2000;71:367S–72.

Houtsmuller UM, van der Beek A. Effects of topical application of fatty acids. Prog Lipid Res. 1981;20:219–24.

Illi S, von Mutius E, Lau S, et al. The natural course of atopic dermatitis from birth to age 7 years and the association with asthma. J Allergy Clin Immunol. 2004;113:925–31.

Imokawa G. A possible mechanism underlying the ceramide deficiency in atopic dermatitis: expression of a deacylase enzyme that cleaves the N-acyl linkage of sphingomyelin and glucosylceramide. J Dermatol Sci. 2009;55:1–9.

Imokawa G, Abe A, Jin K, et al. Decreased level of ceramides in stratum corneum of atopic dermatitis: an etiologic factor in atopic dry skin? J Invest Dermatol. 1991;96:523–6.

Innis SM. Perinatal biochemistry and physiology of long-chain polyunsaturated fatty acids. J Pediatr. 2003;143:S1–8.

Ishibashi M, Arikawa J, Okamoto R, et al. Abnormal expression of the novel epidermal enzyme, glucosylceramide deacylase, and the accumulation of its enzymatic reaction product, glucosylsphingosine, in the skin of patients with atopic dermatitis. Lab Invest. 2003;83:397–408.

Ishikawa J, Narita H, Kondo N, et al. Changes in the ceramide profile of atopic dermatitis patients. J Invest Dermatol. 2010;130(10):2511–4.

Janssens M, van Smeden J, Gooris GS, et al. Lamellar lipid organization and ceramide composition in the stratum corneum of patients with atopic eczema. J Invest Dermatol. 2011 Jun 30. doi:10.1038/jid.2011.175.

Jiang WG, Redfern A, Bryce RP, et al. Peroxisome proliferator activated receptor-gamma (PPAR-gamma) mediates the action of gamma linolenic acid in breast cancer cells. Prostaglandins Leukot Essent Fatty Acids. 2000;62:119–27.

Johansson A, Isolauri E, Salminen S. Alpine currant seed oil as a source of polyunsaturated fatty acids in the treatment of atopic eczema. Budapest: Publishing Company of TUB; 1999.

Jungersted JM, Scheer H, Mempel M, et al. Stratum corneum lipids, skin barrier function and filaggrin mutations in patients with atopic eczema. Allergy. 2010;65:911–8.

Kanehara S, Ohtani T, Uede K, et al. Clinical effects of undershirts coated with borage oil on children with atopic dermatitis: a double-blind, placebo-controlled clinical trial. J Dermatol. 2007a;34:811–5.

Kanehara S, Ohtani T, Uede K, et al. Undershirts coated with borage oil alleviate the symptoms of atopic dermatitis in children. Eur J Dermatol. 2007b;17:448–9.

Kashibuchi N, Hirai Y, O'Goshi K, et al. Three-dimensional analyses of individual corneocytes with atomic force microscope: morphological changes related to age, location and to the pathologic skin conditions. Skin Res Technol. 2002;8:203–11.

Kato A, Fukai K, Oiso N, et al. Association of SPINK5 gene polymorphisms with atopic dermatitis in the Japanese population. Br J Dermatol. 2003;148:665–9.

Kezic S, Kemperman PM, Koster ES, et al. Loss-of-function mutations in the filaggrin gene lead to reduced level of natural moisturizing factor in the stratum corneum. J Invest Dermatol. 2008;128:2117–9.

Kikuchi K, Kobayashi H, O'Goshi K, et al. Impairment of skin barrier function is not inherent in atopic dermatitis patients: a prospective study conducted in newborns. Pediatr Dermatol. 2006;23:109–13.

Kitz R, Rose MA, Schonborn H, et al. Impact of early dietary gamma-linolenic acid supplementation on atopic eczema in infancy. Pediatr Allergy Immunol. 2006;17:112–7.

Komatsu N, Suga Y, Saijoh K, et al. Elevated human tissue kallikrein levels in the stratum corneum and serum of peeling skin syndrome-type B patients suggests an over-desquamation of corneocytes. J Invest Dermatol. 2006;126:2338–42.

Komatsu N, Saijoh K, Kuk C, et al. Human tissue kallikrein expression in the stratum corneum and serum of atopic dermatitis patients. Exp Dermatol. 2007;16:513–9.

Krahn-Bertil E, Hazane-Puch F, Lassel T, et al. Skin moisturization by dermonutrition. In: Rawlings AV, Leyden JJ, editors. Skin moisturization. 2nd ed. 2009. p. 411–25. Chapter 27.

Leung DY. Atopic dermatitis: new insights and opportunities for therapeutic intervention. J Allergy Clin Immunol. 2000;105:860–76.

Linnamaa P, Savolainen J, Koulu L, et al. Blackcurrant seed oil for prevention of atopic dermatitis in newborns: a randomized, double-blind, placebo-controlled trial. Clin Exp Allergy. 2010;40:1247–55.

Macheleidt O, Kaiser HW, Sandhoff K. Deficiency of epidermal protein-bound omega-hydroxyceramides in atopic dermatitis. J Invest Dermatol. 2002;119:166–73.

Man MQ, Hatano Y, Lee SH, et al. Characterization of a hapten-induced, murine model with multiple features of atopic dermatitis: structural, immunologic, and biochemical changes following single versus multiple oxazolone challenges. J Invest Dermatol. 2008;128:79–86.

Manku MS, Horrobin DF, Morse N, et al. Reduced levels of prostaglandin precursors in the blood of atopic patients: defective delta-6-desaturase function as a biochemical basis for atopy. Prostaglandins Leukot Med. 1982;9:615–28.

Marquardt A, Stohr H, White K, et al. CDNA cloning, genomic structure, and chromosomal localization of three members of the human fatty acid desaturase family. Genomics. 2000;66:175–83.

Matsumoto M, Umemoto N, Sugiura H, et al. Difference in ceramide composition between "dry" and "normal" skin in patients with atopic dermatitis. Acta Derm Venereol. 1999;79:246–7.

McMahon A, Butovich IA, Mata NL, et al. Retinal pathology and skin barrier defect in mice carrying a stargardt disease-3 mutation in elongase of very long chain fatty acids-4. Mol Vis. 2007;13:258–72.

Melnik B, Hollmann J, Plewig G. Decreased stratum corneum ceramides in atopic individuals–a pathobiochemical factor in xerosis? Br J Dermatol. 1988;119:547–9.

Muggli R. Systemic evening primrose oil improves the biophysical skin parameters of healthy adults. Int J Cosmet Sci. 2005;27:243–9.

Nakamura MT, Nara TY. Structure, function, and dietary regulation of delta6, delta5, and delta9 desaturases. Annu Rev Nutr. 2004;24:345–76.

Nicollier M, Massengo T, Remy-Martin JP, et al. Free fatty acids and fatty acids of triacylglycerols in normal and hyperkeratotic human stratum corneum. J Invest Dermatol. 1986;87:68–71.

Nikolovski J, Stamatas GN, Kollias N, et al. Barrier function and water-holding and transport properties of infant stratum corneum are different from adult and continue to develop through the first year of life. J Invest Dermatol. 2008;128:1728–36.

Okamoto R, Arikawa J, Ishibashi M, et al. Sphingosylphosphorylcholine is upregulated in the stratum corneum of patients with atopic dermatitis. J Lipid Res. 2003;44:93–102.

Park WJ, Kothapalli KS, Lawrence P, et al. An alternate pathway to long-chain polyunsaturates: the FADS2 gene product Delta8-desaturates 20:2n-6 and 20:3n-3. J Lipid Res. 2009;50:1195–202.

Pilgram GS, Vissers DC, van der Meulen H, et al. Aberrant lipid organization in stratum corneum of patients with atopic dermatitis and lamellar ichthyosis. J Invest Dermatol. 2001;117:710–7.

Prottey C. Investigation of functions of essential fatty acids in the skin. Br J Dermatol. 1977;97:29–38.

Puch F, Samson-Villeger S, Guyonnet D, et al. Consumption of functional fermented milk containing borage oil, green tea and vitamin E enhances skin barrier function. Exp Dermatol. 2008;17:668–74.

Rawlings AV. Trends in stratum corneum research and the management of dry skin conditions. Int J Cosmet Sci. 2003;25:63–95.

Rawlings AV. Recent advances in skin 'barrier' research. J Pharm Pharmacol. 2010;62:671–7.

Rawlings AV, Krahn-Bertil F, et al. Skin moisturisation by dermonutrition: outcomes of a dairy matrix product. 2011;32–43.

Rawlings AV, Matts PJ. Stratum corneum moisturization at the molecular level: an update in relation to the dry skin cycle. J Invest Dermatol. 2005;124:1099–110.

Rawlings AV, Scott IR, Harding CR, et al. Stratum corneum moisturization at the molecular level. J Invest Dermatol. 1994;103:731–41.

Redoules D, Tarroux R, Assalit MF, et al. Characterisation and assay of five enzymatic activities in the stratum corneum using tape-strippings. Skin Pharmacol Appl Skin Physiol. 1999; 12:182–92.

Rogers J, Harding C, Mayo A, et al. Stratum corneum lipids: the effect of ageing and the seasons. Arch Dermatol Res. 1996;288:765–70.

Saaf AM, Tengvall-Linder M, Chang HY, et al. Global expression profiling in atopic eczema reveals reciprocal expression of inflammatory and lipid genes. PLoS One. 2008;3:e4017.

Schaeffer L, Gohlke H, Muller M, et al. Common genetic variants of the FADS1 FADS2 gene cluster and their reconstructed haplotypes are associated with the fatty acid composition in phospholipids. Hum Mol Genet. 2006;15:1745–56.

Schafer L, Kragballe K. Abnormalities in epidermal lipid metabolism in patients with atopic dermatitis. J Invest Dermatol. 1991;96:10–5.

Scharschmidt TC, Man MQ, Hatano Y, et al. Filaggrin deficiency confers a paracellular barrier abnormality that reduces inflammatory thresholds to irritants and haptens. J Allergy Clin Immunol. 2009;124(496–506):e1–6.

Senapati S, Banerjee S, Gangopadhyay DN. Evening primrose oil is effective in atopic dermatitis: a randomized placebo-controlled trial. Indian J Dermatol Venereol Leprol. 2008;74:447–52.

Sprecher H, Luthria DL, Mohammed BS, et al. Reevaluation of the pathways for the biosynthesis of polyunsaturated fatty acids. J Lipid Res. 1995;36:2471–7.

Tagami H, Kobayashi H, O'Goshi K, et al. Atopic xerosis: employment of noninvasive biophysical instrumentation for the functional analyses of the mildly abnormal stratum corneum and for the efficacy assessment of skin care products. J Cosmet Dermatol. 2006;5:140–9.

Taieb A. Hypothesis: from epidermal barrier dysfunction to atopic disorders. Contact Dermatitis. 1999;41:177–80.

Tarroux R, Assalit MF, Licu D, et al. Variability of enzyme markers during clinical regression of atopic dermatitis. Skin Pharmacol Appl Skin Physiol. 2002;15:55–62.

Tokura Y. Extrinsic and intrinsic types of atopic dermatitis. J Dermatol Sci. 2010;58:1–7.

Tsang M, Guy RH. Effect of aqueous cream BP on human stratum corneum in vivo. Br J Dermatol. 2010;163(5):954–8. doi:10.1111/j.1365-2133.2010.09954.x.

Uchida Y, Holleran WM. Omega-O-acylceramide, a lipid essential for mammalian survival. J Dermatol Sci. 2008;51:77–87.

van Gool CJ, Thijs C, Henquet CJ, et al. Gamma-linolenic acid supplementation for prophylaxis of atopic dermatitis–a randomized controlled trial in infants at high familial risk. Am J Clin Nutr. 2003;77:943–51.

van Gool CJ, Zeegers MP, Thijs C. Oral essential fatty acid supplementation in atopic dermatitis-a meta-analysis of placebo-controlled trials. Br J Dermatol. 2004;150:728–40.

Vasilopoulos Y, Cork MJ, Murphy R, et al. Genetic association between an AACC insertion in the 3'UTR of the stratum corneum chymotryptic enzyme gene and atopic dermatitis. J Invest Dermatol. 2004;123:62–6.

Vasireddy V, Uchida Y, Salem Jr N, et al. Loss of functional ELOVL4 depletes very long-chain fatty acids (> or =C28) and the unique omega-O-acylceramides in skin leading to neonatal death. Hum Mol Genet. 2007;16:471–82.

Voegeli R, Doppler S, Joller P, et al. Increased mass levels of certain serine proteases in the stratum corneum in acute eczematous atopic skin. Int J Cosmet Sci. 2011 Jun 24. doi: 10.1111/j. 1468-2494.2011.00671.x.

Voegeli R, Rawlings AV, Breternitz M, et al. Increased stratum corneum serine protease activity in acute eczematous atopic skin. Br J Dermatol. 2009;161:70–7.

Wertz PW, Downing DT. Metabolism of linoleic acid in porcine epidermis. J Lipid Res. 1990;31:1839–44.

White MI, Jenkinson DM, Lloyd DH. The effect of washing on the thickness of the stratum corneum in normal and atopic individuals. Br J Dermatol. 1987;116:525–30.

Yamamoto A, Serizawa S, Ito M, et al. Stratum corneum lipid abnormalities in atopic dermatitis. Arch Dermatol Res. 1991;283:219–23.

Ziboh VA. Essential fatty acids/eicosanoid biosynthesis in the skin: biological significance. Proc Soc Exp Biol Med. 1994;205:1–11.

Ziboh VA, Chapkin RS. Biologic significance of polyunsaturated fatty acids in the skin. Arch Dermatol. 1987;123:1686a–90.

Ziboh VA, Miller CC. Essential fatty acids and polyunsaturated fatty acids: significance in cutaneous biology. Annu Rev Nutr. 1990;10:433–50.

Ziboh VA, Cho Y, Mani I, et al. Biological significance of essential fatty acids/prostanoids/ lipoxygenase-derived monohydroxy fatty acids in the skin. Arch Pharm Res. 2002;25:747–58.

Chapter 12
Hair Biology and Nutritional Influences

Michael Anthonavage

Core Messages

- Hair biology is closely related to skin health and in many instances correlates well with it.
- Hair biology is affected by many factors and is useful for indicating how well the body copes with nutritional, psychological, and pathological stress over periods of time.
- Hair fiber structure, follicle health, follicle cycling, and pigmentation can be examined individually or as a whole when assessing hair health as it relates to nutritional status.

12.1 Introduction

Human hair emanating from the hair follicle is considered a biological tissue composed primarily of keratins, which serve the body in a variety of ways (Robbins 2002). The hair follicle itself is an amazingly complex skin appendage that to this day still warrants intense investigation because it involves so many aspects of biological science including embryology, cell biology, molecular biology, stem cell biology, tissue engineering, and nutritional biology (Stenn and Paus 2001). Healthy hair is certainly an indicator of one's general well-being, particularly during the reproductive years, but hair loss itself is not a life-threatening event. Hair loss, graying hair, or unwanted hair growth does, however, affect the quality of life of the individual by altering his or her self-esteem and confidence, which has both emotional and sociological consequences.

M. Anthonavage (✉)
Presperse, LLC, 635 Pierce Street, Somerset, NJ 08873, USA
e-mail: manthonavage@presperse.com

A. Pappas (ed.), *Nutrition and Skin: Lessons for Anti-Aging, Beauty and Healthy Skin*,
DOI 10.1007/978-1-4419-7967-4_12, © Springer Science+Business Media, LLC 2011

Hair integrity, growth and function rely on balance, as do all biological systems. That balance is maintained largely in part by the homeostatic mechanisms of survival of those cells and tissues that comprise the organ. A nutritional component therefore is the first line on inquiry in any living system and thus the focus of this chapter.

In this review, the relationship with nutritional status and food components is explored in regard to the health and growth of human hair. Although skin and hair are generally spoken about in a similar fashion, an attempt is made to make a distinction between the two because other chapters in this book address in detail the attributes of skin and its appendages as they relate to nutrition. It should be also noted that in humans, there are surprisingly few scientific studies correlating nutrition and hair health because the models for hair growth and health are primarily other species of mammals, such as mice and primates (Randall et al. 2003). Many attempts to grow hair ex vivo are underway, and it has been deemed quite difficult, time-consuming, and costly to say the least. Encouraging results have been demonstrated using cell culture modeling of particular aspects of hair follicle biology; but, as can be said about most models, they have their shortcomings in terms of dimensionality and utility to consumer-perceived benefits. This, then, leads us to investigate epidemiological studies and disease states to correlate nutritional aspects as they relate to hair health (Hambridge 2003).

12.2 Healthy Hair

Healthy hair is easy to recognize and is generally full bodied, shiny, lustrous, and free of flakes and damage. Unhealthy hair can result from overprocessing and, more importantly, inadequate nutrition or disease processes. Although graying hair or loss of hair as a result of aging is not considered a pathological process, the lack of color or hair loss does counter one's perception of healthy hair and is somewhat connected to environmental exposure to oxidants and ultraviolet (UV) radiation (Lu et al. 2009). To understand how vital nutrients are to healthy hair, a quick synopsis of hair biology is in order to ground the discussion.

12.3 Hair Biology

The average human scalp has approximately 100,000 hairs emanating from hair follicles. Hair is one of the fastest growing tissues in the body. In addition to having a decorative role in society, hair has a variety of biological functions, including protection from UV radiation and insects. It acts as a sensory tool and conduit for excretion of sebaceous oils. Hair is also important for insulation and warmth in terms of thermoregulation and has been shown to harbor stem cell populations and immune cells, which are important for the regenerative capacity of the skin and innate immunity, respectively (Table 12.1; Stenn and Paus 2001). Hair follicles undergo a continuous cycle to maintain the presence of hair outside the body.

Table 12.1 Functions of the hair shaft and follicle (Stenn and Paus 2001)

Functions of the hair shaft
• Decoration, social communication, and camouflage
• Protects against trauma and insect penetration
• Protects against electromagnetic radiation
• Provides a sensory "antenna" to feel the environment
• Insulates against heat loss and heat gain
• Mechanism of cleansing skin surface of squames, dirt, and parasites
• Mechanism of outward transport of environmental signals: sebum, pheromones
Functions of pilosebaceous follicle
• Produces and moors the shaft
• Provides epithelial and dermal reservoir for normal renewal and reparative response
• Provides sensory apparatus for detecting shaft movement
• Provides melanocyte reservoir for shaft pigmentation and epidermal repigmentation
• Produces and releases sebum for shaft processing and epidermal surface protection
• Provides reservoir of Langerhans cells

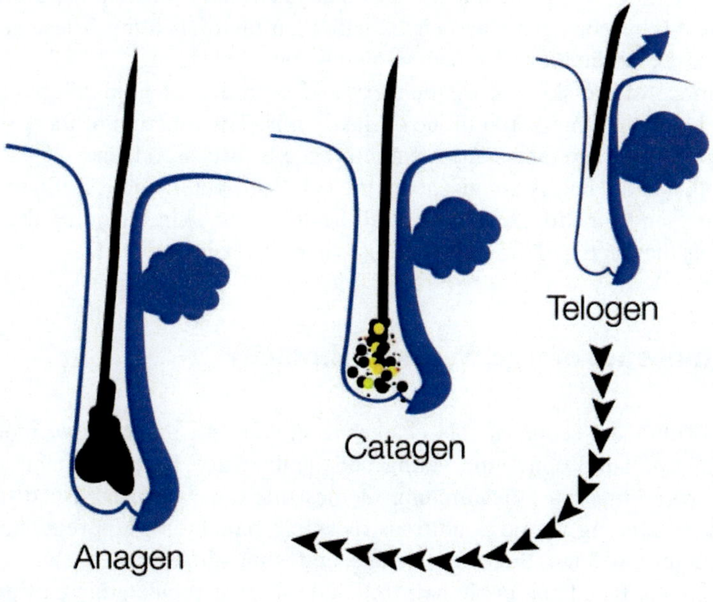

Fig. 12.1 Hair cycle

The hair continuously cycles through three primary phases and at least two subphases of growth (Fig. 12.1). Of the primary phases, anagen is considered the period of growth; catagen is the regressing phase; and telogen is the resting or quiescent phase. Each of these phases has several morphologically and histologically distinguishable subphases. For a comprehensive overview of the biology of cycling hair and its controls see Stenn and Paus (Stenn and Paus 2001). Prior to the start of cycling is a subphase of follicular morphogenesis (formation of the follicle), and at the end of the cycle there is a shedding phase, or exogen, that is independent of the anagen and telogen phases. Normally in the human scalp up to 90% of the hair follicles are in the anagen phase, 10–14% are in the telogen phase, and 1–2% are in the catagen phase. The length of the cycle can vary on different parts of the body. For example, eyebrows cycle in approximately 4 months, whereas it takes the scalp 3–4 years to complete a cycle. It is for this reason that eyebrow hairs have a much shorter length than hairs on the head.

Connected to each hair follicle is an afferent sensory nerve ending that allows the hair to be utilized for sensory perception. In addition, each follicle is supplied with its own source of blood from the capillaries in the dermis. This blood supply is the primary source of nutrients to the hair bulb located at the base of the hair follicle where growth is believed to be initiated and sustained. The outside of the hair follicle is not vascularized and receives nutrients from surrounding interstitial spaces through diffusion from local capillary beds in the dermis. These spaces are drained and perfused by a series of lymphatic vessels. The importance of all these vessels ensures a continuous rate of perfusion of both nutrients and waste that ultimately affect the heath of the follicular tissue. Finally, virtually all terminal hairs on the body have a sebaceous gland associated with their microanatomy. These glands are involved in the production of sebum, which is funneled through the follicular canal to the surface of the skin for the delivery and excretion of lipid-soluble products from the blood and interstitial tissue of the dermis. The function of these materials is widely debated, and their nutritional relevance is discussed later.

Overall, healthy hair is not just about the hair itself but includes follicular health, sebaceous gland function, and the overall health of the skin including the neurons and vessels that feed and drain the dermis where the hair is rooted.

12.4 Influence of Age, Sex, and Ethnicity

It is not within the scope of this chapter to explore all the possible modulatory mechanisms of hair production, cycling, and maintenance; however, mention should be made regarding some overarching themes that are well established to give a broad understanding of the conditions in which hair biology is regulated under normal conditions. First, there are sex, age, and ethnic differences in the production of hair. In men, facial and pubic hair follicle biology is regulated in part by androgens, and these changes are considered some of the hallmarks of puberty. Hormonal changes in the physical properties of hair have also been noted in postmenopausal women where the ratio of estrogens to androgens is altered resulting in thicker more noticeable facial hair (Wines 2001). With age, the most noticeable effects are the loss of hair in men and loss of pigment in both sexes. Additionally, upon close observation, there are distinct phenotypic differences in the thickness and shape of hair fibers among the different ethnic groups (Wolfram 2003). For the most part, however, these differences are not determined by diet but by genetics, and the differences do play a role in the detection of trace elements, metals, and toxins acquired through ingestion.

12.5 Influence of Adipose Tissue

During anagen, the hair follicle is at its longest, stretching deep into the dermal layer of the skin. Located deep within the skin beyond the dermis is a layer of adipose tissue called the hypodermis. Here, adipocytes rich in triglycerides are the most predominant cell type. Fat tissue is highly vascularized and hormonally active as it responds quickly to excess caloric intake and a sedentary lifestyle. Adipose tissue is also a source of a variety of inflammatory mediators such as tumor necrosis factor-α (TNFα) and matrix metalloproteinases (MMPs), which can ultimately affect homeostasis of the surrounding tissue. Of particular interest here is the fact that the skin's subcutaneous fat layer diminishes with age (Sepe 2010). Fat cells are associated with large, healthy hair follicles and may be associated with providing nutritional support to the follicles (Rei 2006). Also in support, subcutaneous fat is considered a substantial stem cell reservoir (Rei 2006). This role of fat has been associated with hair-promoting effects in mice (Won 2010).

12.6 Apoptosis and Hair Health

Apoptosis is a nonpathological process of cellular death brought on by a wide array of conditions, one of which is mitochondrial insufficiency. A mosaic of respiratory chain deficiencies in a subset of cells in various tissues (e.g., heart, skeletal muscle, colonic crypts, neurons) is typically found in aged humans (Bratic 2010). Because apoptosis plays such a critical role in the hair cycle (catagen) and apoptosis can be triggered by a variety of inflammatory events, it is deemed an important biological process in the maintenance of healthy hair. Ensuring proper cellular respiration through the presence of antioxidant and anti-inflammatory compounds at all levels of the body, including the skin and its hair follicles, is prudent.

12.7 Stress

Stress plays a critical role in hair biology, particularly transient hair loss. Physiological stress—trauma, high fever, chronic disease—is a well-known cause of a type of hair loss called alopecia effluvium, which results in the loss of hair 2–3 months after the incident (Kligman 1961). Not uncommon is the loss of hair postpartum (telogen gravidarum) (Kligman 1961). Emotional stress has long been discussed controversially as a cause of hair loss. However, solid proof of stress-induced hair growth inhibition is still missing. If psychoemotional stress does affect hair growth, it would be mediated via definable neurorendocrine and/or neuroimmunological signaling pathway(s) (Peters 2006). Positive support through adequate diet and sleep are fundamental mediators of stress.

12.8 Circulation and Perfusion of Nutrients

The rate of blood flow through the skin is variable in the human body owing to its primary role in regulating body temperature, which is based on both internal metabolism and external temperature. Because tissues in the skin are fed via the circulation, blood flow becomes an important issue when considering the dose of nutrients. With skin being one of the largest organs in the body and the fact that it is the primary thermoregulatory organ, there are a considerable number of vessels dedicated to perfusion of blood and nutrients. Supplying sufficient nutrients to microorgans such as the hair follicle and sebaceous glands are as important as acquiring the nutrients. The long and winding road to the skin and hair starts with ingestion, digestion, and absorption of nutrients. Hair growth and maintenance are in part linked to vitamin B and folate assimilation through the gastrointestinal tract. Hair growth potential is optimum when specific parameters for biochemical variables such as red blood cell (erythrocyte) generation are in place. These include erythrocyte and serum folate concentrations within the normal range, serum vitamin B_{12} levels of 300–1,000 ng/l, hemoglobin levels >13.0 g/dl, and serum ferritin concentrations of ≥70 ng/ml (Rushton 1993). One must also consider the skin- and tissue-specific metabolic alterations to the nutrient(s) supplied along with their half-lives in the surrounding tissues when contemplating delivery of nutrients. Permeation to the appropriate cell types in the follicle and uptake by receptor-specific moieties are essential provided the tissue and its cells are functioning correctly. Finally, degradation of nutrients as well as their metabolic waste products needs to be considered. Other factors affecting nutrient bioavailability are associated with external stresses such as UV radiation and smoke, pharmaceutical drug use, and pathologies associated with the hair shaft and skin

12.9 Hair Fiber Stress

As the external portion of the hair is a keratin fiber structure susceptible to external effects—whether mechanical, physical, or chemical—excessive sun exposure and heat from processing are the most frequent causes of hair shaft structural impairment (Sebetic 2008). Photochemical impairment of the hair includes degradation and loss of hair proteins as well as degradation of melanin in the hair. Ultraviolet B (UVB) radiation is responsible for hair protein loss, and UVA radiation is responsible for color changes (Sebetic 2008). Absorption of radiation in photosensitive amino acids of the hair and their photochemical degradation produces free radicals and changes the spectral absorbance of the hair (Noqueira 2006). Free radicals have a particular adverse effect on hair proteins, especially keratin (Sebetic 2008). The total amount and type of amino acids present are more susceptible to photodegradation (tryptophan, cysteine, tyrosine, histidine) depending on the hair type. Male hairs have more cysteine than female hairs; and usually dark hairs have more cysteine than light hairs (Robbins 2002). Other extrinsic factors include smoke and pollution, both of which increase the level of oxidative stress.

12.10 Influence of Pharmaceutical Drugs and Herbal Extracts

Pharmaceutical drugs, chemotherapy, and the components of medicinally used herbals can cause unhealthy hair formation and hair loss. One of the most troublesome in terms of emotional distress is the shedding of hair due to chemotherapy. When an arrest of mitotic activity occurs in the follicle, many interacting factors influence and promote the shedding pattern. Oral contraceptives and hormone replacement therapy, progestin, and progesterone, respectively, have been documented to cause hair loss as have retinoids and angiotensin-converting enzyme (ACE) inhibitors and androgens (Fiedler 2003; Tosti and Pazzaglia 2007). Anticoagulant components of plants such as those that inspired the production of warfarin and coumadin and the popular antidepressive St. John's wort have also been associated with sporadic hair loss (Umlas 1988; Cornbleet 1957; Parker 2001). As a result, much attention should be paid to the secondary metabolites of plants when considering therapeutic application of plant extracts. It should be noted that *Cuscuta reflexa,Citrullus colocynthis*, and *Eclipta alba* are traditional herbs known for their ability to grow hair when topically applied (Prendiville 1992). However, there is scant evidence of effective oral therapy rooted in natural substances. Compounding the problem is the fact that there are limited studies available for documenting such approaches. When such studies are available, they are usually performed in murine models, which have not been shown to correlate well with human hair growth or hair maintenance.

12.11 Disease and Inflammation

Many disease states associated with abnormal skin and hair follicle biology focus on immunity, infection, and inflammation. A careful examination of shed hair reveals the etiology of a variety of alopecias due to systemic immune processes. Telogen effluvium is a condition preceded by severe systemic stress occurring at least 2 months prior to the loss of normal hair.

Psoriasis is a chronic autoimmune disease affecting the skin resulting in thickened scaly patches of affected areas that may include the scalp. There are many theories as to what causes psoriasis, but two areas of involvement seems to be suspect: an inappropriate immune response and the interaction between epithelial cells, called keratinocytes, in the epidermis. Keratinocytes are derived from stem cells that originate from the epidermal compartment and/or cells from within the bulge area of the follicle (Schreder 2010). Of interest is the idea that psoriasis may have its own hair phenotype. It has been demonstrated that there are more dysplastic hair shafts in psoriatic plaques than in skin without psoriasis (Stanimirovic 1998). The reason for this is still up for debate. Mechanical disfigurement of the follicular shaft may be caused by thickening of the skin, which then changes the way hairs extrude from the skin. Others speculate that aberrant differentiation in the cells that make up the hair shaft are affected. To the latter point, a common treatment for psoriasis is the use of cholecalciferol (vitamin D_3), which has antiproliferative and prodifferentiation effects on skin, thus restoring the balance between keratinocytes (Gregoriou 2010).

Dysregulation of the hair follicle is also noted in a plethora of inflammatory conditions, notably alopecia areata. Although multifactorial, alopecia areata is complete loss of hair in localized areas of the scalp and involves, in part, inflammatory cytokines and stress hormones of the interleukin family, interferon family, and TNFα (Gregoriou 2010). This type of alopecia is now understood to represent an organ-restricted, T-cell-mediated autoimmune disease of hair follicles. Disease induction is associated with collapse of hair follicle immune privilege in humans (Gihar 2007).

Infections on or in the skin usually indicate an imbalance in the skin's innate immunity. When this imbalance is present, opportunistic infection becomes prevalent. Tinea capitis is a fungal infection of the scalp, usually caused by *Microsporum* or *Trichophyton* species of dermatophytes. It usually occurs in prepubertal patients. The most severe form of tinea capitis is a kerion—a fluctuant, boggy lesion with overlying hair loss (Gregoriou 2010). Tinea capitis can result in widespread hair loss with increased fragility of the hairs and frequent breakage, suggesting inappropriate or altered hair fiber synthesis.

12.12 Historical Link Between Hair and Nutrition

Many skin problems originate from extrinsic factors but also have an underlying dietary/nutritional component. The relation between nutrients and skin comes from the incidence of skin problems as a result of nutritional deficiencies. According to maritime records, in 1497 Vasco de Gamma recorded the deaths of 100 of his 160 sailors after a scurvy outbreak during his voyage around the Cape of Good Hope. Noted in the ship's log were observations telling of skin hemorrhage and hair loss (Stewart 1953). Three centuries later, James Lind, a pioneer of naval hygiene, confirmed the link between scurvy and ascorbic acid (Rushton 2002b; Stewart 1953). Daniel Whistler in 1645 is credited with the first published book on the description of vitamin D deficiency in a thesis on rickets; but it was not until 1922 that McCullum et al. demonstrated the link with vitamin D (Rushton 2002b; McCollum 1922). More recently, the beneficial role of vitamin D analogues in psoriasis and skin's innate immunity have been established (Rushton 2002b; Loser 2009). Other examples of dietary components that have an effect on skin are nicotinic acid, riboflavin, thiamine, and pyridoxine. Of particular note is the report that pyridoxine deficiency produces dermatological findings with features similar to that of essential fatty acid deficiency (Witten 1952).

Deficient consumption of several vitamins and essential fatty acids clearly results in cutaneous manifestations. Although the frequency of nutritional deficiencies is low in developed countries, unbalanced and incomplete diets as a result of disease, aging, and chemical substances abuse may influence health and thereby affect the skin (Prendiville 1992; Smith 2000). Therefore, correcting and optimizing the diet may not only prevent skin and hair problems but may also correct any potentially underlying condition. Studies investigating the effects of oral supplementation with relatively high doses of vitamins, trace minerals, and fatty acids have indicated the possibility that dietary factors can modulate skin function and possibly hair health (McLaren 1993; Prendiville 1992).

12.13 Hair Color

Hair color can also be an indicator of nutritional status, especially in cancer patients and children. The incidence of melanoma is higher in fair-skinned than in dark-skinned individuals. There is in vitro evidence to support the notion that fair skin or poorly tanning skin contains more saturated fatty acids and therefore is more subject to UV-induced oxidative stress (Cario-Andre 2005; Bessou-Touva 1998). Additionally, the Western diet, which is rich in omega-6 polyunsaturated fatty acids, is more associated with the development of a particular type of skin cancer called cutaneous melanoma (Cario-Andre 2005). Research has shown that in malnourished children melanin distribution along the shaft of the hair fiber itself is reduced (McKenzie 2007). It is possible that reduced intake or availability of tyrosine, a key substrate in melanin synthesis, plays a role in the reduction of hair melanin content during periods of malnutrition (McKenzie 2007). The precise mechanism(s) by which melanin content is reduced in hair and the role of aromatic amino acid availability in hair color change does warrant further investigation.

12.14 Protein and Amino Acids

The influence of protein on hair follicle biology in humans emanates primarily from studies on energy malnutrition (kwashiorkor and marasmus), particularly in areas of the world where starvation and poor diet are prevalent (Badaloo 2006). Subjects suffering from late stages of human immunodeficiency virus (HIV) infection, anorexia nervosa, or cancer cachexia exhibit profiles of altered protein metabolism leading to manifestations in skin and hair homeostasis (Hediger 2000; Vafaie 2005; Delmore 1997). Hair growth takes place from specific regions in the hair follicle. Dermal tissue projects into the follicle base to form the dermal papilla, where there is a network of capillary blood vessels to supply oxygen, energy, and the amino acids needed for growth (Fig. 12.2). Gas chromatography studies have shown that it takes more than 4 weeks for both essential and nonessential amino acids to be detected in hair using stable isotopic abundance analysis after a group of young women were subjected to additional meat intake versus a group who had had meat removed from their diet (Petzke 2009). The phenotype of a hair root's response to

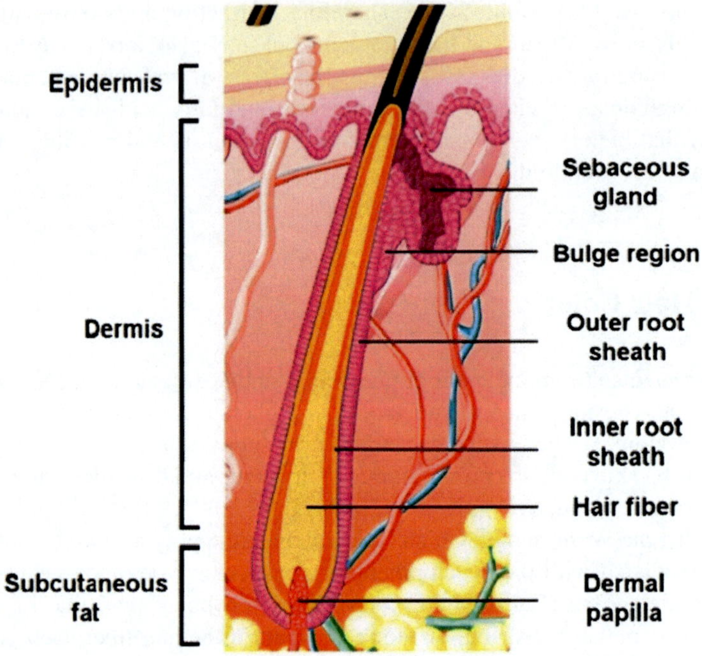

Fig. 12.2 Anatomy of the hair follicle (www.keratin.com/aa/aa007.shtml)

protein malnutrition has been studied widely in children with kwashiorkor and marasmus conditions. It has been documented that the hair is easily epilated and has a finer diameter upon growth. In addition, there are alterations in the production of the hair that make it less curly, depigmented, and fragile to mechanical stress. As for the skin, there are also notable changes in keratinization, pigmentation, and delayed wound healing with aberrant protein intake.

The amino acids proline and lysine along with ascorbic acid from the diet are linked to a healthy dermis, where hair follicles reside. With increased age, dermal proteins (collagen and elastin) begin to degrade. Ascorbic acid is a water-soluble vitamin that acts as a reducing agent required for synthesis of healthy collagen fibers through hydroxylation of proline and lysine. More commonly, it protects the body against damage caused by free radicals. Additionally, suboptimal ingestion of L-lysine was correlated with ferritin deficiency and iron-related hair shedding in women (Rushton 2002a). This highlights the importance of serum ferritin levels in regard to hair loss. In women without systemic inflammation or other underlying disorders, serum ferritin levels ≤30 ng/ml are strongly associated with telogen hair loss (Moeinvaziri 2009).

12.15 Androgens, Insulin Resistance, and Lipids

Healthy hair is not just about preventing hair loss; healthy hair may also play a role in preventing inappropriate or unwanted hair growth, or hirsutism, which is more of a symptom than a disease. The term hirsutism is usually applied to women when hair starts to grow in areas typically seen in men. This condition has been associated with endocrine imbalances associated with androgen metabolism and is commonly influenced by polycystic ovary disease or the onset of menopause.

One of the primary enzymes involved in androgen metabolism is 5α-reductase, which converts testosterone to dihydrotestosterone. The latter is biologically active on androgen-sensitive hair tissue. Interestingly, there are a number of commonly used food/herbal products that have 5α-reductase properties. Green tea and Saw Palmetto are two such products. Green tea catechins have been demonstrated to inhibit testosterone conversion (Hiipakka 2002). 5α-Reductase has two isoforms that are found in two distinct areas of the skin. Isoform type 1 is found in keratinocytes and sebaceous glands, and type 2 is found in hair follicles. Flavonoids from natural products that were potent inhibitors of the type 1 5α-reductase include myricetin, quercitin, baicalein, and fisetin. Biochanin A, daidzein, genistein, and kaempferol are flavonoids that are much better inhibitors of the type 2 isozyme than the type 1 isozyme (Hiipakka 2002).

Insulin resistance has also been linked to hirsutism. Hirsutism is a finding that can lead to subsequent metabolic diagnoses such as the metabolic syndrome. Metabolic syndrome describes a cluster of risk factors associated with being overweight that can lead to an increase in cardiovascular risk and further obesity (Sarac 2007). Because diet and insulin resistance are closely linked, there may be a plausible link to controlling hirsutism; however, more studies are needed in this area.

Control of fat intake as part of a balanced diet is critical to preventing the metabolic syndrome, cardiovascular disease, and obesity, conditions that have been linked to the biology of hair growth. Dietary fats and fatty acids have been shown to modulate skin especially in hyperinflammatory conditions such as eczema and atopic dermatitis (Boelsma 2001). In particular, the consumption of omega-3 and omega-6 fatty acids for the management of inflammation is well documented (Fu et al. 2001).

Skin not only receives lipids from the diet, it also acts as an excretory organ via the production of sebum excreted through the sebaceous canal. Sebaceous glands produce sebum, which is composite of various lipids that are secreted to the surface of the skin via the hair shaft for shaft processing and surface protection (Stenn and Paus 2001). Sebum's lipid content is unique to humans, and it has been demonstrated that essential fatty acids, α-tocopherol, and coenzyme Q_{10} are secreted to the skin through this mechanism, thus reinforcing the link between diet and skin (Passi 2002; Thiele 1999). The action of sebum has been debated over the years, but there is evidence that it protects against oxidative damage and inflammation. Much of the free fatty acid content in sebum is the result of bacterial action on triglycerides as well as the result of delivery from the plasma in the case of linoleic acid and linolenic acid.

Foods and seed extractions rich in these fatty acids are becoming quite popular with the average consumer and may play a role in acne (Pappas 2009). What these lipids have to do with hair formation is still not well understood, but they have recently been associated with cicatricial alopecia (Stenn 2010).

The term "cicatricial alopecia" refers to a diverse group of rare disorders that destroy the hair follicle, replace it with scar tissue, and cause permanent hair loss. A key desaturase found in sebaceous glands, called stearoyl-coenzyme A (CoA) desaturase, is an endoplasmic reticulum-bound enzyme that catalyzes the delta 9-*cis* desaturation of saturated fatty acyl-CoAs, with the preferred substrates being palmitoyl-CoA and stearoyl-CoA (Miyazaki 2003). Deletion of this enzyme in mice has demonstrated abnormal sebaceous gland development and disrupted hair follicle organization (Binczek 2007; Paton 2009). Stearoyl CoA desaturase activity leads to the production of the monounsaturated fatty acids palmitoleate and oleate, both of which are bactericidal against Gram-positive (but not Gram-negative) organisms in vitro (George 2005; Wille and Kydonieus 2003). Fatty acids have been associated with barrier homeostasis via controlled bacterial colonization and keratinocyte differentiation in the epidermis, and links to follicular health have been made (Wille and Kydonieus 2003). Driving these actions are the peroxisome proliferator-activated receptors (PPARs), which are fatty acid-activated transcription factors that belong to the nuclear hormone receptor family. PPARs regulate sebocyte differentiation and promote hair follicle growth through prodifferentiating effects in keratinocytes in normal and inflammatory conditions (Michalik 2007). Owing to their diverse function and activation schemes, PPAR biology depends not only on dietary fatty acids but also on the ligands produced during inflammation. They thus have important effects on hair and sebaceous cell biology.

12.16 Trace Elements

Trace element nutrition is the most studied aspect of hair as it relates to nutrition. Zinc, copper, magnesium, and selenium have been examined for their effects on hair biology. In alopecia areata, serum zinc levels are significantly decreased in subjects who are resistant to standard treatment when compared to control subjects. Of particular note, copper and magnesium levels were elevated in this study but not significantly (Bhat 2009).

Zinc is an essential micronutrient for human metabolism; it is a cofactor for more than 100 enzymes, facilitates protein folding, and helps regulate gene expression. Zinc is one of the oldest medicines in recorded history, with its use having been reported on 3000-year-old papyrus. Zinc is also one of the most studied trace elements in hair, as many have looked to hair analysis for a quick and noninvasive way to evaluate body zinc indexes. Patients with malnutrition, alcoholism, inflammatory bowel disease, and malabsorption syndromes are at increased risk of zinc deficiency. Manifestations of zinc deficiency in the skin include nail dystrophies and hair abnomalities (Saper 2009). It should be noted that zinc level determination is controversial, and many factors play into an inaccurate zinc level determination in hair—mainly the use of zinc pyrithione in shampoos. Other factors include age, sex, and geographic location of the individual (Nelder et al. 1991).

Additional studies examining the role of trace elements in alopecia found that there are statistically significant differences in the serum copper content in alopecia areata and alopecia universalis patients (Mussalo-Rauhamaa 1986). Copper is the key mineral in an enzyme called lysyl oxidase, which participates in the crosslinking of elastin fibers in the dermis. A combination of collagen and elastin is essential for tissues such as blood vessels and for a functional matrix in the dermis. Mice with a spontaneous recessive mutation in the lysyl oxidase gene show growth retardation, cyclic and progressive hair loss, hyperplastic epidermis, abnormal hair follicles, cardiac muscle degeneration, and reduced amounts of collagen and elastin in the skin and heart (Hyashi 2004).

Selenium along with zinc has been getting much attention in the cold and flu market as they are said to reduce the symptoms of the common cold. It should be noted that with most trace elements megadoses inhibit the very biology they are supposed to assist. Thus is the case with selenium. High doses of ingested selenium have been documented to have adverse effects in natural killer cell activity, hepatotoxicity, gastrointestinal disturbances, and nail and hair loss (Vinceti 2001).

12.17 Lipophilic Vitamins

Key lipophilic vitamins to health are β-carotene, cholecalciferol, and tocopherol. β-Carotene is cleaved to form retinal, which in turn gives rise to retinol and retinoic acid, both of which play vital roles in epidermal biology including cell proliferation and epidermal differentiation. In vitro and in vivo studies have demonstrated that of the retinoids tested those most effective in altering the levels of cellular retinoic acid-binding protein in the skin were also capable of significantly altering hair cycle dynamics. There appears to be a relation between the ability of retinoids to increase cellular retinoic acid-binding protein, increase ^{3}H-thymidine incorporation, and alter the dynamics of the hair cycle (Bazzano 1993). In addition, β-carotene concentrations in plasma have been shown to be significantly lower in patients with alopecia than in control subjects (Naziroglu 2000). These data provide evidence for a potential role of increased lipid peroxidation, lipid metabolism, and decreased lipophilic antioxidants in alopecia (Naziroglu 2000).

High doses of serum tocopherol did not prevent chemotherapy-induced alopecia (Martin-Jimenez 1986). This is a rather surprising finding in light of the fact that tocopherol is secreted via sebum from the sebaceous gland and into the hair canal (Thiele 1999).

Another lipophilic vitamin, cholecalciferol, is of nutritional origin but can also be generated in the skin by UV light. After conversion by enzymes in both the liver and kidneys, cholecalciferol avidly binds and activates the vitamin D receptor, thereby regulating a large number of genes involved in skin homeostasis. In skin, vitamin D receptor expression in keratinocytes is essential for maintenance of the normal hair cycle (Bouillon 2006; Demay 2007). It is noteworthy that vitamin D receptor knockout mice express a hair follicle cycling defect and a hyperproliferative phenotype resulting in disordered skin structure, epidermal thickening, and alopecia (Dowd et al. 2010). Therefore, the vitamin D receptor and not serum cholecalciferol deficiency may be the driver in specific types of alopecia (Bouillon 2006).

12.18 Hydrophilic Vitamins

The hydrophilic vitamins, such as the family of B vitamins, have been shown to be important for normal skin homeostasis. Cyanocobalamin, or B_{12}, occurs in two metabolically active forms: (1) methylcobalamin, which is linked to DNA, protein, and lipid metabolism, and (2) coenzyme B_{12}, which has a role in carbohydrate and fat metabolism. Cyanocobalamin deficiency is associated with a reduction in glutathione, which normally inhibits tyrosinase. Tyrosinase activity is fundamental to the production of melanin in the skin and hair (Mori 2001). To demonstrate the pigmentation effects of cyanocobalamin, a case of reversible generalized hyperpigmentation of the skin and nails with reversible premature gray hair due to vitamin B_{12} deficiency has been reported. The cause of the vitamin B_{12} deficiency in these patients was considered to be due to pernicious anemia. In these studies, the pigmentation of his skin and hair returned to normal after treatment with intramuscular cyanocobalamin (Noppakun 1986).

Biotin is another water-soluble vitamin that acts as an essential cofactor for five carboxylase enzymes, each of which catalyzes an essential step in intermediary metabolism. In addition, biotin-dependent carboxylase catalyzes the fixation of bicarbonate in organic acids and plays a crucial role in the metabolism of fatty acids, amino acids, and glucose. Carboxylase activities decrease substantially in response to biotin deficiency. Deficiency here may be caused by insufficient dietary uptake of biotin, drug–vitamin interactions, and perhaps an increase in biotin catabolism during pregnancy and in smokers. Symptoms of biotin deficiency in skin include dermatitis and hair loss (Zempleni 2008). Biotin is one of the most widely used vitamins for hair as it is said to promote healthy hair growth and protect against dryness presumably through long-chain fatty acid incorporation into the cuticle, but sufficient evidence is lacking. It also increases the elasticity of the hair's cortex, preventing breakage. One reason to believe in this action comes from the effect of biotin on fatty acid metabolism. Biotin is a cofactor for acetyl-CoA carboxylase, which catalyzes the rate-limiting step in fatty acid elongation (Zempleni 2008). In infants, children, and adults, deficiency of biotin causes alopecia and a characteristic scaly, erythematous dermatitis distributed around body orifices (Mock 1991). The rash closely resembles that of zinc deficiency (Mock 1991). There is evidence that impaired fatty acid metabolism secondary to reduced activities of the biotin-dependent carboxylases (especially acetyl-CoA carboxylase) plays an etiological role in the dermatological manifestations of biotin deficiency (Mock 1991).

12.19 Tools for Studying Hair Loss

Certain tools are best suited for diagnosis in medical practice, whereas others can only be used to monitor hair growth under treatment in clinical arenas. The techniques can be classified as either invasive (e.g., biopsies in scarring alopecia), semiinvasive (e.g., trichography), or noninvasive (e.g., global hair counts, phototrichography, electron microscopy, laser scanning microscopy) (Hillmann 2009). Each has its pros and cons and should be combined with accurate nutritional assays in the serum to correlate the cause and effect of the underlying problem in the hair and its follicle.

12.20 Final Thoughts

Malnutrition results from a deficiency of one or more basic nutrients and may be caused by insufficient dietary intake, malabsorption, poor utilization of nutrients, and/or an increase in metabolic breakdown or catabolism, each of which can contribute to poor hair health in some form or fashion. A range of clinical and metabolic changes occur in the skin, particularly the hair follicles, as a result of generalized abnormalities at the cellular level. Although some signs are characteristic of a specific nutrient deficiency, others may overlap with skin symptoms and thus confuse the clinician. This being said, multiple deficiency states should be considered when evaluating hair health as it relates to nutritional status (Prendiville 1992).

To establish the cause of aberrant hair biology, one requires a history to identify known triggers, biochemical investigations to exclude endocrine, nutritional or autoimmune etiologies, and in many cases histology to identify the earliest stages of alopecia. The duration of hair loss upon presentation helps predict those patients in whom further investigation will have the greatest yield. Dietary influences on hair cycling and healthy hair growth are documented; and such events as caloric restriction and chronic starvation have been shown to cause hair shedding (Harrision 2009). Conversely, a common practice by many looking to correct nutritional deficiencies is to supplement for what is believed to be a loss of critical nutrients. Excessive intake of nutritional supplements may indeed cause hair loss and are not recommended in the absence of a proven deficiency (McLaren 1993). Although nutritional factors affect the hair directly, they also affect the skin and its functions. In the management of subjects with hair loss, one should also include any problems with excessive scaling or xerotic (dry) skin, as these conditions influence follicle health (Rushton 2002b). According to Montagna (Harison 2002):

> It is unfortunate that baldness and grey hair has been approached with an eye towards commercialism. Locked within the metamorphosing hair follicles in the balding scalp are all the secrets of growth and differentiation. Searching for these secrets should transcend the eagerness to "re-grow" hair on a bald scalp, an achievement which is of no great consequence. When we know these answers, we shall have the key, not to hair growth alone, but to all growth, which is, after all, the basis of all biological phenomena.

References

Badaloo AV. Lipid kinetic differences between children with kwashiorkor and those with marasmus. Am J Clin Nutr. 2006;83(6):1283–8.

Bazzano G. Effects of retinoids on follicular cells. J Invest Dermatol. 1993;101(1):138S–42.

Bessou-Touva S. Chimeric human epidermal reconstructs to study the role of melanocytes and keratinocytes in pigmentation and photoprotection. J Invest Dermatol. 1998;111(6):1103–8.

Bhat YJ. Trace element levels in alopecia areata. Indian J Dermatol Venereol Leprol. 2009;75(1):29–31.

Binczek E. Obesity resistance of the stearoyl-CoA desaturase-deficient (scd1–/–) mouse results from disruption of the epidermal lipid barrier and adaptive thermoregulation. Biol Chem. 2007;388(4):405–18.

Boelsma E. Nutritional skin care: health effects of micronutrients and fatty acids. Am J Clin Nutr. 2001;73:853–64.

Bouillon R. Vitamin D resistance. Best Pract Res Clin Endocrinol Metab. 2006;20(4):627–45.

Bratic I. Mitochondrial energy metabolism and aging. Biochim Biophys Acta. 2010;1997 (6–7):961–7. Epub 2010 Jan 11.

Cario-Andre M. Polyunsaturated fatty acids partially reproduce the role of melanocytes in the epidermal melanin unit. Exp Dermatol. 2005;14(3):194–201.

Cornbleet T. Alopecia from coumarin. AMA Arch Derm. 1957;75(3):440–1.

Delmore G. Assessment of nutritional status in cancer patients: widely neglected? Support Care Cancer. 1997;5(5):376–80.

Demay M. Role of the vitamin D receptor in hair follicle biology. J Steroid Biochem Mol Biol. 2007;103(3–5):344–6.

Dowd DR. The 1,25-dihydroxyvitamin D3-independent actions of the vitamin D receptor in skin. Steroid Biochem Mol Biol. 2010;121(1–2):317–21. Epub 2010 April 11.

Fiedler VC. Diffuse alopecia: telogen hair loss (Chapter 10). In: Olsen EA, editor. Disorders of hair growth: diagnosis and treatment. 2nd ed. New York: McGraw-Hill; 2003. p. 303–32.

Fu Z, Attar-Bashi NM, Sinclair AJ. 1-14C-linoleic acid distribution in various tissue lipids of guinea pigs following an oral dose. Lipids. 2001;36(3):255–60.

George P. A toll-like receptor 2-responsive lipid effector pathway protects mammals against skin infections with gram-positive bacteria. Infect Immun. 2005;73(8):4512–21.

Gihar A. Lymphocytes, neuropepetides, and genes involved in alopecia areata. J Clin Invest. 2007;117(8):2019–27.

Gregoriou S. Cytokines and other mediators in alopecia areata. Mediators inflamm. 2010;2010 928030. Epub 2010 March 11.

Hambridge M. Biomarkers of nutritional exposure and nutritional status. J Nutr. 2003;133(8):948s–55.

Harison S. Telogen effluvium. Clin Exp Dermatol. 2002;27(5):389–95.

Harrision S. Diffuse hair loss: its triggers and management. Cleve Clin J Med. 2009;76(6):361–7.

Hediger C. Cutaneous manifestation in anorexia nervosa. Schweiz Med Wochenschr. 2000;130(16):565–75.

Hiipakka RA. Structure-activity relationships for inhibition of human 5alpha-reductases by polyphenols. Biochem Pharmacol. 2002;63(6):1165–76.

Hillmann K. Diangnosis of hair disorders. Semin Cutan Med Surg. 2009;28(1):33–8.

Hyashi K. Progressive hair loss and myocardial degeneration in rough coat mice: reduced lysyl oxidase-like (LOXL) in the skin and heart. J Invest Dermatol. 2004;123(5):864–71.

Kligman AM. Pathologic dynamics of human hair loss I. Telogen effluvium. Arch Dermatol. 1961;83:175–98.

Loser K. Regulation of cutaneous immunity by the environment: an important role for UV irradiation and vitamin D. Int Immunopharmacol. 2009;9(5):587–9.

Lu Z et al. Profiling the response of human hair follicles to ultraviolet radiation. J Invest Dermatol. 2009;129(7):1790–804. Epub 2009 Jan 22.

Martin-Jimenez M. Failure of high-dose topherolto prevent alopecia induced by doxorubicin. N Engl J Med. 1986;315(14):894–5.
McCollum EV. Studies on experimental rickets. XXI. An experimental demonstration of the existence of a vitamin which promotes calcium deposition. J Biol Chem. 1922;53:293–312.
McKenzie CA. Childhood malnutrition is associated with a reduction in the total melanin content of scalp hair. Br J Nutr. 2007;98(1):159–64.
McLaren DS. Fat soluble vitamins. In: Garrow JS, James WPT, editors. Human nutrition, dietetics. 9th ed. Edinburgh: Churchill Livingstone; 1993. p. 208–38.
Michalik L. Peroxisome proliferator-activated receptors (PPARs) in skin health, repair and disease. Biochim Biophys Acta. 2007;1771(8):991–8.
Miyazaki M. Role of stearoyl-coenzyme a desaturase in lipid metabolism. Prostaglandins Leukot Essent Fatty Acids. 2003;68(2):113–21.
Mock DM. Skin manifestations of biotin deficiency. Semin Dermatol. 1991;10(4):296–302.
Moeinvaziri M. Iron status in diffuse telogen hair loss among women. Acta Dermatovenerol Croat. 2009;17(4):279–84.
Mori K. Generalized hyperpigmentation of the skin Due to vitamin B12 deficiency. J Dermatol. 2001;28(5):282–5.
Mussalo-Rauhamaa H. Element concentrations in serum,erythrocytes, hair and urine of alopecia patients. Acta Derm Venereol. 1986;66(2):103–9.
Naziroglu M. Antioxidants and lipid peroxidation status in the blood of patients with alopecia. Cell Biochem Funct. 2000;18(3):169–73.
Nelder K. The biochemistry and physiology of zinc metabolism (Chapter 50). In: Goldsmith L, editor. The biochemistry and molecular biology of the skin. 2nd ed. New York: Oxford University Press; 1991. p. 1329–50.
Noppakun N. Reversible hyperpigmentation of skin and nails with white hair Due to vitamin B12 deficiency. Arch Dermatol. 1986;122(8):896–9.
Noqueira AC. About photo-damage of human hair. Photochem Photobiol Sci. 2006;5(2):165–9.
Pappas A. The relationship of diet and acne. Dermatoendocrinol. 2009;1(5):262–7.
Parker V. Adverse reactions to St John's Wort. Can J Psychiatry. 2001;46(1):77–9.
Passi S. Lipophilic antioxidants in human sebum and aging. Free Radic Res. 2002;36(4):471–7.
Paton CM. Biochemical and physiological function of stearoyl-CoA desaturase. Am J Physiol Endrinol Metab. 2009;297(1):E28–37.
Peters EM. Hair growth and inhibition by psychoemotional stress: a mouse model for neural mechanisms in hair growth control. Exp Dermatol. 2006;15:1–13.
Petzke KJ. Hair protein and amino acid ^{13}C and ^{15}N abundances take more than 4 weeks to clearly prove influences of animal protein intake in young women with a habitual daily protein consumption of more than 1g per kg body weight. Rapid Commun Mass Spectrom. 2009;23(16):2411–20.
Prendiville JS. Skin signs of nutritional disorders. Semin Dermatol. 1992;11:88–97.
Randall V, Sundberg JP, Philpott M. Animal and in-vitro models for the study of hair follicles. J Investig Dermatol Symp Proc. 2003;8:39–45.
Rei O. The importance of adipose derived stem cells and vascularized tissue regeneration I the field of tissue transplantation. Curr Stem Cell Res Ther. 2006;1:13–20.
Robbins CR. Chemical and physical behavior of human hair. 4th ed. New York: Springer; 2002.
Rushton DH. Management of hair loss in women. Dermatol Clin. 1993;11(1):47–53.
Rushton DH. Causes of hair loss and the developments in hair rejuvenation. Int J Cosmet Sci. 2002a;24(1):17–23.
Rushton DH. Nutritional factors and hair loss. Clin Exp Dermatol. 2002b;27(5):396–404.
Saper RB. Zinc: an essential micronutrient. Am Fam Physician. 2009;79(9):768–72.
Sarac F. Assessment of insulin resistance in the idiopathic hirsutism. Gynecol Obstet Invest. 2007;63(3):126–31.
Schreder A. Facing towards the epidermal stem cells (review). Int J Mol Med. 2010;26(2):171–4.
Sebetic K. UV damage of the hair. Coll Antropol. 2008;32(2):163–5.

Sepe A. Aging and regional differences in fat cell progenitors-A mini-review. Genetol. 2011;57(1): 66–75. Epub 2010 Jan 29.
Smith KE. Cutaneous manifestations of alcohol abuse. J Am Acad Dermatol. 2000;43:1–16.
Stanimirovic A. Has psoriasis its own characteristic trichogram? J Dermatol Sci. 1998;17(2):156–9.
Stenn KS. Lipids t the Top of hair biology. J Invest Dermatol. 2010;130(5):1205–7.
Stenn KS, Paus R. Controls of hair follicle cycling. Physiol Rev. 2001;81(1):449–94.
Stewart CP. Treatise on scurvy. Edinburgh: Edinburgh University Press; 1953. p. 145–8.
Thiele JJ. Sebaceous gland secretion is a major physiologic route of vitamin E delivery to skin. J Invest Dermatol. 1999;113(6):1006–10.
Tosti A, Pazzaglia M. Drug reactions affecting hair: diagnosis. Dermatol Clin. 2007;25:223–31.
Umlas J. Warfarin-induced alopecia. Cutis. 1988;42(1):63–4.
Vafaie J. Alopecia in association with sexually transmitted disease: a review. Cutis. 2005;76(6):361–6.
Vinceti M. Adverse health effects of selenium in humans. Rev Environ Health. 2001;16(4):233–51.
Wille JJ, Kydonieus A. Palmitoleic acid isomer (C16:1delta6) in human skin sebum is effective against gram-positive bacteria. Skin Pharmacol Appl Skin Physiol. 2003;16(3):176–87.
Wines N. Menopause and the skin. Australas J Dermatol. 2001;42(3):149–60.
Witten P. Polyethenoid fatty acid metabolism VI, effect of pyridoxine on essential fatty acid conversion. Arch Biochem Biophys. 1952;41:266.
Wolfram LI. Human hair: a unique physiochemical composite. J Am Acad Dermatol. 2003;48(6):s106–14.
Won CH. Hair growth promoting effects of adipose tissue-derived stem cells. J Dermatol Sci. 2010;57(2):134–7.
Zempleni J. Biotin and biotinidase deficiency. Expert Rev Endocrinol Metab. 2008;3(6):715–24.

Chapter 13
Detecting and Monitoring Nutrients on Skin Using Noninvasive Methods

Georgios N. Stamatas and Nikiforos Kollias

Core Messages

- Due to metabolic processes inside the body, most nutrients cannot be detected directly on skin, and only a few might be detected in either their original molecular form or as one of their metabolic products.
- Based on their Raman spectral fingerprint, lycopene and β-carotene can be detected noninvasively on skin. Their apparent concentration increases with a carotenoid-rich diet.
- Nutritional lipid uptake can be detected on the lips using infrared spectroscopy.
- Immediate and long-term effects of nutrition may be quantitatively evaluated using noninvasive optical methods.

13.1 Introduction

Technological progress in optics and electronics during the last couple of decades has allowed the development of sensitive analytical methods. Although originally such methods were designed for analysis of small samples, many quickly were adapted with probes that could come in contact with a living tissue. In this way, one is able to acquire information about the biochemical composition of the investigated tissue without the need to perform a biopsy. Researchers have applied such methods on human and animal skin in vivo and have acquired a plethora of information about the skin and subcutaneous tissue (Kollias and Stamatas 2002).

G.N. Stamatas (✉)
Johnson & Johnson Consumer France,
1, rue Camille Desmoulins, Issy-les-Moulineaux 92787, France
e-mail: gstamata@its.jnj.com

A. Pappas (ed.), *Nutrition and Skin: Lessons for Anti-Aging, Beauty and Healthy Skin*,
DOI 10.1007/978-1-4419-7967-4_13, © Springer Science+Business Media, LLC 2011

In this chapter we discuss how optical noninvasive methods can be applied to detect and monitor the evolution of concentrations of nutrients on the skin and the acute and long-term effects of nutrients that can be manifested as alterations in skin physiological parameters. Because of catabolic processes during digestion, the nutrient components (primarily proteins, polysaccharides, and lipids) are quickly broken down to their constituents (amino acids, monosaccharides, free fatty acids) in the gastrointestinal tract or other metabolic organs (e.g., liver) before entering the peripheral vessels of the bloodstream, which then delivers these molecules to tissues such as the skin. Therefore, it can be challenging to detect primary nutrients in the skin. There are, however, some interesting exceptions that we examine here. Finally, digestion of certain nutrients may affect skin physiology indirectly—for example, by activating the immune system and eliciting allergic reactions that can manifest as skin rashes. In that case, optical noninvasive methods can be employed to monitor the evolution of the reaction quantitatively.

When trying to detect nutrients or their effects in the skin, we need to consider how they might arrive there or why we might expect them or their metabolites to localize in the skin. For lipophilic molecules, we might expect to find an increased concentration in the stratum corneum (SC) or in subcutaneous fat, each of which presents challenges for detection: The first is very thin, and the second is difficult to explore because of the limited penetration depth by light into the skin. These problems might be worked out by exploring techniques that would allow probing of large areas of the SC to increase the path length through the absorber or by using near-infrared (IR) radiation to probe to greater depths, reaching the subcutaneous tissue or probing on skin sites where the overlying layers are thinner, such as the area under the eye.

Microspectroscopy of histological sections might prove useful for determining the site in the tissue where the chromophores might concentrate. The simplistic approach of considering the skin as a "cuvette" where the chromophores are uniformly distributed is naive and has led to minimal use of optical techniques to detect and monitor the time evolution of chromophore concentrations. Edwards and Duntley (1939) showed that the distribution of chromophores is not constant over the skin surface area but varies. Therefore, when designing suitable detection methods, we need to consider such variations in regions of high or low concentration of the molecule of interest. As an example, carotenoids concentrate in areas of the body where the SC is thick (palm of the hand, sole of the foot), and in these same spots the concentration of melanin is the lowest, minimizing the interference of signals.

13.2 Short Description of Optical Noninvasive Methods

13.2.1 Brief Introduction to Skin Optics

The methods we discuss in this chapter are based on interactions of electromagnetic radiation (light) in the ultraviolet (UV) B (280–315 nm), UVA (315–400 nm), visible (400–700 nm), and IR (700 nm–300 μm) parts of the spectrum. The fundamental

Table 13.1 Skin conditions and characteristics that can be documented in vivo with optical non-invasive methods; physical principles

Clinical relevance	Skin component involved	Method	Physical principle
Skin pigmentation	Melanin and deoxygenated hemoglobin	DRS, orthogonal polarization imaging, spectral imaging	Light absorption and light scattering
Erythema	Oxygenated hemoglobin	DRS, orthogonal polarization imaging, spectral imaging	Light absorption and light scattering
Edema	Skin water content	Spectral imaging	Light absorption and light scattering
Epidermal proliferation	Tryptophan	Fluorescence spectroscopy	Fluorescence
Collagen and elastin glycation	Collagen and elastin cross-links	Fluorescence spectroscopy	Fluorescence
Skin composition	Sebaceous lipids, SC lipids, SC water	ATR-FTIR	Energy absorption by vibrational modes of molecular bonds
Skin composition	Water, keratin, lipids, NMF components, etc.	Raman spectroscopy	Raman scattering
Skin microstructure	Cells, skin layers, etc.	Confocal reflectance microscopy	Light scattering

DRS diffuse reflectance spectroscopy, *SC* stratum corneum, *ATR-FTIR* attenuated total reflection–Fourier transform infrared spectroscopy, *NMF* natural moisturization factor

principle is that light is allowed to interact with the tissue and is then collected and analyzed. More specifically, an incident beam of light with a defined intensity and spectral composition is directed toward the surface of the skin site of interest. Skin is translucent, which means that light can penetrate its surface and can travel through the tissue. Light interacts with the components of the skin in ways that can alter: (1) its intensity (light absorption by skin chromophores); (2) its spectral composition (fluorescence, Raman scattering, higher-order effects such as second harmonic generation and multiphoton fluorescence); and (3) its direction of travel (Mie and Rayleigh scattering, fluorescence, Raman scattering, higher order effects). Together with light absorption, light scattering is by far the most frequent phenomenon. Following multiple scattering events, part of the light is remitted back out of the skin tissue. This light can be collected and analyzed for its intensity and spectral composition. Table 13.1 summarizes the optical noninvasive methods used on skin and the physical phenomena (light–tissue interactions) on which they are based.

An exception to the above fundamental principle is the method of skin chemiluminescence (CL) analysis. This method is used to quantify the amount of light that spontaneously is emitted from the skin during oxidation reactions and therefore does not require any incident light.

Table 13.2 Necessary components of the optical noninvasive methods for in vivo skin measurements

Method	Light source	Probe	Detector
DRS	LED, incandescent lamp, etc.	Bifurcated fiberoptic randomized probe	Spectrometer
Fluorescence spectroscopy	Xe arc lamp, laser, etc.	Bifurcated fiberoptic randomized probe	Photomultiplier, spectrometer, etc.
ATR-FTIR	Laser	ATR crystal	Photomultiplier
Raman spectrometer	Laser or LEDs	Lens	Photomultiplier, spectrometer, etc.
Raman confocal microspectroscopy	Laser	Microscope lens	Photomultiplier, spectrometer, etc.
Reflectance confocal microscopy	Laser	Microscope lens	Photodiode
Imaging	Flash or continuous light source	Lenses and filters	Digital camera

LED light emitting diode

13.2.2 Spectroscopic Methods

The spectroscopic methods usually involve a light source, a means of delivering the light to the skin and collecting the remitted light (commonly a fiberoptic probe or a microscope lens), and a spectrometer that analyzes the collected light. Table 13.2 summarizes the components for the most commonly used spectroscopic methods.

For diffuse reflectance spectroscopy (DRS) a broad band of visible light is used. The intensity spectral profile of the remitted light can be analyzed to give apparent concentrations of the skin chromophores (primarily melanin and hemoglobins) and an indication of the light-scattering strength of the tissue. Note that we use the term "apparent" concentration because an actual concentration of a molecule can only be defined for a homogeneous solution, whereas skin tissue is inherently inhomogeneous from the macroscopic to the molecular level. This method can be used to evaluate skin pigmentation, erythema (redness), and blanching reactions (Kollias and Baqer 1988; Kollias et al. 1995, 2001; Zonios et al. 2001; Stamatas et al. 2008b).

Fluorescence spectroscopy commonly utilizes a monochromatic (e.g., laser) or narrow-band source (e.g., xenon arc lamp filtered at the appropriate band). The remitted light is often analyzed using a monochromator and a photomultiplier. The excitation (source) light is always at shorter wavelengths than the emission (remitted). In skin, tryptophan and other aromatic amino acids have characteristic fluorescence signals that can be detected. Particularly tryptophan fluorescence has been used as a marker of epidermal cell proliferation (Gillies et al. 2000; Doukas et al. 2001) and has been shown to be age-dependent (Tian et al. 2001; Stamatas et al. 2006a). Collagen and elastin crosslinks in the dermis also demonstrate characteristic fluorescence patterns. The fluorescence of collagen crosslinks has been related to skin aging (Tian et al. 2001; Stamatas et al. 2006a) and accumulation of advanced glycation end-products (AGE) in diabetes (Monnier et al. 2005).

Attenuated total reflection–Fourier transform infrared (ATR-FTIR) spectroscopy is used to analyze absorptions in the mid-IR part of the spectrum, indicating the presence of specific chemical bond vibrations. Identification of these bonds can provide information about concentration and molecular orientation of molecules that contain these bonds. For example, spectral analysis of the absorption due to the amide bond confers information about proteins and ceramides. The ATR method limits the interrogated tissue volume to a couple of micrometers, which in skin corresponds to the uppermost layers of the SC. ATR-FTIR has been used in the analysis of SC lipids and secreted sebaceous lipids (Brancaleon et al. 2000).

Similar to ATR-FTIR, Raman scattering depends on interactions of electromagnetic energy with molecular bond vibrations. In contrast to ATR-FTIR, in Raman there is a shift in the wavelength of the incident light that can be either positive (Stokes) or negative (anti-Stokes). The extracted information is often complementary to that acquired with ATR-FTIR. Raman spectrometers have been developed using focusing lenses that result in a relatively large illumination area on the skin (2 mm circular disk), and therefore the collected information is the average over the illuminated tissue volume (Hata et al. 2000). In another configuration Raman spectroscopy can be coupled to a microscope objective in a confocal arrangement that allows focusing the light beam on a small volume ($1 \times 1 \times 5$ mm). By moving the objective lens vertically, one can move the sampling volume to different depths in the tissue. Raman confocal microspectroscopy (RCM) can thus be applied to extract chemical information at sequential depths. This information can in turn be used to construct concentration profiles of the skin constituents (Caspers et al. 2000). Raman spectroscopy and RCM have been used to quantify water and components of the natural moisturization factor (NMF) in the SC (Caspers et al. 2001) as well as topically applied substances (Caspers et al. 2002; Pudney et al. 2007; Stamatas et al. 2008a) and accumulation of carotenoids (e.g., following dietary supplementation) (Hata et al. 2000).

Chemiluminescence is the phenomenon of spontaneous light emission from the skin due to oxidative reactions and recombination of free radicals. CL has been used to study the oxidative stress on the skin following UVA exposure and the effects of antioxidants (Sauermann et al. 1999; Ou-Yang et al. 2004).

13.2.3 Macroscopic and Microscopic Imaging Methods

The principles of spectroscopy can be applied during macroscopic or microscopic imaging. Here the light source is a flash unit or a continuous source, and the detector is a two-dimensional sensor array that may record information in the form of a digital image. An added advantage of imaging is that many instances it may be without contact.

Using polarization filters in front of the source and the camera lens in an orthogonal orientation, one can exclude the specular reflections and collect purely color information of the imaged object (Phillips et al. 1997). In the case of skin imaging, color is the result of light absorption by the skin chromophores, primarily melanin in the epidermis and hemoglobins in the vascular network.

By filtering the source or the detector with a sequence of narrow bands, one can collect a stack of images that can be used to contract maps of chromophore concentrations of a skin area (Stamatas et al. 2006b; Stamatas and Kollias 2007). This method is termed spectral imaging and is the imaging equivalent of DRS.

The source light and the detected light can be independently filtered to spectral bands specific to skin molecules that can fluoresce. As with fluorescence spectroscopy, fluorescence imaging can be used to monitor the extent of collagen crosslinking, but it can also be used to accentuate pigmentation issues (Kollias et al. 1997).

To achieve higher magnification, the principles of orthogonal polarization imaging, spectral imaging, and fluorescence imaging can be applied to video microscopy of skin.

13.3 Direct Detection of Nutrients in Skin

13.3.1 Carotenoids

Carotenoids are a family of polyene molecules with 40 carbon atoms. They cannot be synthesized in the body and can only be introduced to it in the diet. They are thought to be potent antioxidants and free radical quenchers, and they therefore play a protective role for the tissues where they are found. Carotenoids have been shown to inhibit carcinoma formation in animal models (Bernstein et al. 1998). Owing to their lipophilic nature they naturally accumulate to relatively high concentrations in the SC, where they are thought to play a protective role from sun-induced oxidative stress. Their accumulation in the SC also makes them suitable targets for noninvasive detection.

Lycopene and β-carotene absorb light in a broad, short visible range that is demonstrated by an orange color. Although reflectance spectroscopy has been used to estimate the carotenoid level in the skin of fruit (Ruiz et al. 2008; Lopez-Sanchez et al. 2010) or fish (Grether et al. 2005), to our knowledge there has been limited application of this method on human skin (Prince and Frisoli 1993; Stahl et al. 1998). This could is due to its low specificity for carotenoid detection in the presence of much higher concentrations of melanin and hemoglobins. More specifically, the β-carotene absorption maxima are broad and occur in the same wavelength range (400–500 nm) in which melanin absorbs strongly and hemoglobins exhibit strong Soret bands. A more important reason for the difficulty of detecting carotenoids by DRS is that the available path length for carotenoid absorption is extremely short (limited to the thickness of the SC), and therefore high concentrations are required for detection. Some carotenoids also exhibit fluorescence, although with low quantum yields (Gillbro and Cogdell 1989); therefore, fluorescence detection has not been exploited toward that end.

Their long carbon chain and in particular the many double bonds in carotenes gives them a relatively strong Raman cross section in the region of C–C and C=C stretch vibrations (around 1,160 and 1,525 cm^{-1}, respectively) (Hata et al. 2000). These signals can be followed by noninvasive Raman microspectroscopy (Darvin et al. 2009) and can even be enhanced when the excitation source is in an area where the molecule of interest absorbs light strongly, a method termed resonance Raman. In fact, using the appropriate wavelengths, we can enrich the contribution to the measured Raman signal of a particular carotenoid of interest (Ermakov et al. 2004).

In the past, the resonance Raman method was applied to detect carotenoids on skin, and it had been used for this purpose on the shell of shell fish (Salares et al. 1977) and in the human eye (Bernstein et al. 1998). In 2000, Hata and coworkers used an 488-nm argon laser line as a source to excite the Raman bands of carotenoids on ex vivo and in vivo skin samples (Hata et al. 2000). This method measures primarily lycopene and β-carotene. They reported a good intersubject reproducibility but large intrasubject variability in measured carotenoid concentrations. They also reported site variability, with the palm and the forehead showing higher values than the volar forearm and the dorsal hand. A large portion of the measured carotenoids were shown to be localized in the SC. Comparing a small number of patients, they showed that healthy volunteers ($n=6$) had higher carotenoid levels than patients with either actinic keratosis ($n=14$) or basal cell carcinoma ($n=14$). It was not clear if there was a casual relation. Moreover, the apparent concentrations of skin carotenoids using resonance Raman were related to fruit and vegetable intake (Ermakov et al. 2005; Rerksuppaphol and Rerksuppaphol 2006; Darvin et al. 2008) and inversely related to sun exposure and smoking (Ermakov et al. 2005). Apparently a diet rich in "ecological" eggs was able to increase the skin carotenoids detected with this method by 20% (Hesterberg et al. 2009), although this study was small ($n=6$) and with no control group. In another study, 25 volunteers following a 4-week lycopene-deprived diet received oral lactolycopene or placebo for 12 weeks. Apparent concentrations of lycopene on skin were sensitive to lycopene deprivation and supplementation (Blume-Peytavi et al. 2009).

Recently Bergeson et al. reported the development of an instrument for evaluating the amount of skin carotenoids using light-emitting diodes (LEDs) and multichannel detection using four photomultipliers with single-photon sensitivity (Bergeson et al. 2008). The replacement of the argon laser with LEDs makes this instrument compact and easily portable. These investigators showed good correlation between the readings of the LED and the laser-based instruments.

The resonance Raman method was criticized for lack of good calibration as a comparison of the Raman data with traditional high-performance liquid chromatography (HPLC) (invasive and requires sample destruction) of the same samples was inconclusive (Hammond and Wooten 2004). However, more recent data show good correlation between Raman-based carotenoid readings and blood plasma levels (Ermakov et al. 2005; Zidichouski et al. 2009). It was also criticized for not taking into account the effect of various optical properties of the skin tissue among skin sites and among individuals (e.g., due to variations in chromophore concentration, epidermal morphology) (Hammond and Wooten 2005). A method where the Raman signal is normalized to an independent signal may therefore be more accurate (Darvin et al. 2009).

Using the dual wavelength approach, Ermakov et al. used resonance Raman to detect carotenoid signals in an attempt to separately detect lycopene and β-carotene in skin (Ermakov et al. 2004). They showed that at each excitation wavelength (488.0 and 514.5 nm) the Raman signal has contributions from both carotenes. Provided that the relative molecular contributions to these signals are known, the relative apparent concentrations of these two molecules can then be calculated. It was reported that the relative concentration ratios of β-carotene to lycopene varied significantly among subjects.

13.3.2 Nutrient Lipids

Owing to metabolic processes, most injected molecules undergo significant changes in their structure before they can reach skin tissue. A recent report by Yoshida and coworkers claimed direct measurement of dietary fatty acids on human lips using ATR-FTIR (Yoshida et al. 2009). The lips were chosen as an area unaffected by sebaceous activity that could mask changes in lipid composition following ingestion of fatty acids. A derivation of the cis-alkene (HC=CH) vibration band (3,005–3,015 cm^{-1}) was used to follow changes in polyunsaturated fatty acid concentration over 8 h following a docosahexaenoic acid (DHA)-containing triglyceride breakfast or control. The lips were washed before measurement to remove any food material. The investigators showed that the apparent concentration of DHA could be followed in a time-dependent manner that followed the same kinetics when the concentrations were established by extraction and thin-layer chromatography (TLC) analysis. Hence, these investigators showed that ATR-FTIR is an excellent choice for detecting dietary lipids in the SC noninvasively.

13.4 Detection and Monitoring Nutrient Effects on Skin

What we eat can have measurable effects on skin properties. These effects can be immediate, such as in the case of food allergies; mid term, such as in the case of antioxidant activity following ingestion of carotenoids; or long term, such as the chronic accumulation of advanced glycation end-products (AGEs) in the extracellular matrix of the dermis. In the following paragraphs, we discuss the use of noninvasive methods for monitoring changes in skin properties following ingestion of a particular nutrient or a particular long-term diet.

13.4.1 Immediate Effects of Nutrition on Skin: Food Allergies

One of the most common immediate symptoms of food allergy are skin erythema (redness), angioedema (swelling), and utricaria (itchy skin). The most common delayed symptoms involving skin include exacerbation or worsening of eczema (Werfel 2001). Erythema and edema are skin conditions that can be readily documented, quantified, and monitored over the period of the manifestation and its resolution or over the period of treatment. These conditions alter the optical properties of the skin in the visible and near-IR part of the spectrum correspondingly and can therefore be evaluated with spectroscopic and imaging methods.

It has been shown that the concentration of oxygenated hemoglobin, determined by analysis of diffusely reflected spectra from the skin, is a single parameter that can be used to quantify skin erythema, as in the case of irritant contact dermatitis (Kollias et al. 1995) or UV-induced inflammation (Kollias and Baqer 1988; Kollias et al. 2001).

Whereas the DRS method is based on point measurements with a probe that needs to come in contact with the skin, spectral imaging is a noncontact method that can be used to generate concentration maps of the skin chromophores, including oxygenated hemoglobin (Stamatas and Kollias 2007). These maps contain the added information of the spatial extent of erythema. It has been demonstrated that the sensitivity of the oxygenated hemoglobin map is better than visual inspection, and therefore the method can be used to document subclinical levels of erythema (Groner et al. 1999; Stamatas and Kollias 2007; Sprigle et al. 2009).

Spectral imaging has also been used to document and quantify the extent of edema (Stamatas et al. 2006b). This is achieved by generating a water concentration map based on the light-absorbing properties of water in the near-IR. Note that local accumulation of fluid (mostly water) is what causes skin swelling in edema. In a histamine iontophoresis protocol, a typical wheel-and-flare reaction includes both erythema and edema and can resemble the skin hives that are observed during a food allergy reaction. This protocol was used to demonstrate that both skin reactions can be documented simultaneously using spectral imaging.

Finally, orthogonal polarization imaging can be a simpler, more cost-effective alternative to spectral imaging with regard to documenting erythema, but not edema. Orthogonal polarization imaging has been used to document erythema in acne lesions (Phillips et al. 1997). Compared to spectral imaging, orthogonal polarization is more qualitative than quantitative.

13.4.2 Midterm Effects of Nutrition on Skin: Antioxidant Activity and Photoprotection

Measuring spontaneously emitted photons from the skin surface in a completely dark room can provide an evaluation of the oxidative status of the skin (Sauermann et al. 1999). Following UVA exposure, for example, the CL counts initially increase and then decay in an exponential-like fashion. We have reported that the initial burst of the CL signal depends on the UVA fluence rate, whereas the decay of the signal following exposure can be related to the UVA dose (Ou-Yang et al. 2004). This method has been used to demonstrate that topical application of vitamin C before exposure significantly can reduce the UVA-induced CL signal (Ou-Yang et al. 2004; Hagens et al. 2008). It is conceivable, then, that a diet rich in antioxidants confers a similar effect.

Using clinical evaluation and DRS for assessing UVB-induced skin erythema, it has been demonstrated that oral administration of tomato paste that is rich in lycopene may provide protection against acute and potentially longer-term aspects of photodamage (Rizwan et al. 2010).

13.4.3 Long-Term Effects of Nutrition on Skin: Glycation

Long-lived proteins, such as collagen and elastin molecules in the dermis, in the presence of sugars undergo a series of nonenzymatic glycation and oxidation reactions that result in intra- and intermolecular crosslinks. It has been shown that such AGEs accumulate in the dermis as a function of chronological aging (Odetti et al. 1992; Stamatas et al. 2006a). Moreover, it has been demonstrated that the rate of AGE accumulation is higher in diabetic patients (Dyer et al. 1993; Ediger et al. 2009), which is thought to be the result of chronically high levels of glucose in the blood.

A convenient way to measure certain dermal AGEs, in particular pentosidine and *N*,ε-(carboxymethyl) lysine, is in vivo fluorescence spectroscopy (Odetti et al. 1992; Hull et al. 2004). The fluorescence excitation/emission maxima are 335/385 nm for pentosidine and 370/440 nm for *N*,ε-(carboxymethyl) lysine. Using fluorescence spectroscopy, we and others have confirmed that AGEs accumulate in the skin with aging (Odetti et al. 1992; Stamatas et al. 2006a; Corstjens et al. 2008). Moreover, this method has been used to demonstrate that increased levels of dermal AGEs correlate with increasing risk of developing diabetes-related health complications (Hartog et al. 2005; Meerwaldt et al. 2005; Lutgers et al. 2006; Monami et al. 2008; Chabroux et al. 2010; Conway et al. 2010) and can even be a strong predictor of cardiac mortality in diabetic patients (Meerwaldt et al. 2007).

Although it is well accepted that AGEs increase with aging and are in higher levels in diabetic patients, it remains controversial whether a diet high in carbohydrates would accelerate the accumulation of AGEs in dermal collagen (Cefalu et al. 1995; Novelli et al. 1998; Lingelbach et al. 2000). On the other hand, there have been reports that certain nutrients may prevent high rates of dermal AGE formation in animals (Sajithlal et al. 1998; Rutter et al. 2003; Thirunavukkarasu et al. 2004; Nandhini et al. 2005). In an animal model suitable for genetic studies, *Caenorhabditis elegans*, we have shown, using in vivo fluorescence spectroscopy, that dietary restriction and metabolic mutations can affect the formation of AGE-like pigments (Gerstbrein et al. 2005).

Interestingly, AGE-related skin fluorescence has been shown to increase transiently postprandially (Stirban et al. 2008), which opens the possibility of its potential use in metabolic monitoring.

13.5 Conclusion

Technological advancements in electronics and photonics catalyzed the development of noninvasive methods for in vivo analysis of skin composition. We reviewed the scientific literature regarding application of such methods to monitor the presence of molecules on the skin that come directly from diet and measurable dietary effects on skin properties.

Take-Home Messages

- Particular carotenes can be monitored in the skin by Raman spectroscopy.
- Specific dietary lipids can be detected on skin sites lacking sebaceous activity (lips) using ATR-FTIR.
- Skin manifestations of food allergies can be accurately recorded and monitored using reflectance spectroscopy, polarization imaging, and spectral imaging.
- Long-term effects of a carbohydrate-rich diet or an inability to metabolize carbohydrates fully, as in diabetes, result in AGEs that can be monitored using fluorescence spectroscopy.

References

Bergeson SD, Peatross JB, et al. Resonance Raman measurements of carotenoids using light-emitting diodes. J Biomed Opt. 2008;13(4):044026.

Bernstein PS, Yoshida MD, et al. Raman detection of macular carotenoid pigments in intact human retina. Invest Ophthalmol Vis Sci. 1998;39(11):2003–11.

Blume-Peytavi U, Rolland A, et al. Cutaneous lycopene and beta-carotene levels measured by resonance Raman spectroscopy: high reliability and sensitivity to oral lactolycopene deprivation and supplementation. Eur J Pharm Biopharm. 2009;73(1):187–94. Epub 2009 May 13.

Brancaleon L, Bamberg MP, et al. Spectral differences between stratum corneum and sebaceous molecular components in the mid-IR. Appl Spectrosc. 2000;54(8):1175–82.

Caspers PJ, Lucassen GW, et al. Automated depth-scanning confocal Raman microspectrometer for rapid in vivo determination of water concentration profiles in human skin. J Raman Spectrosc. 2000;31:813–8.

Caspers PJ, Lucassen GW, et al. In vivo confocal Raman microspectroscopy of the skin: noninvasive determination of molecular concentration profiles. J Invest Dermatol. 2001; 116(3):434–42.

Caspers PJ, Williams AC, et al. Monitoring the penetration enhancer dimethyl sulfoxide in human stratum corneum in vivo by confocal Raman spectroscopy. Pharm Res. 2002;19(10):1577–80.

Cefalu WT, Bell-Farrow AD, et al. Caloric restriction decreases age-dependent accumulation of the glycoxidation products, N epsilon-(carboxymethyl)lysine and pentosidine, in rat skin collagen. J Gerontol A Biol Sci Med Sci. 1995;50(6):B337–41.

Chabroux S, Canoui-Poitrine F, et al. Advanced glycation end products assessed by skin autofluorescence in type 1 diabetics are associated with nephropathy, but not retinopathy. Diabetes Metab. 2010;36(2):152–7. Epub 2010 Feb 6.

Conway B, Edmundowicz D, et al. Skin fluorescence correlates strongly with coronary artery calcification severity in type 1 diabetes. Diabetes Technol Ther. 2010;12(5):339–45.

Corstjens H, Dicanio D, et al. Glycation associated skin autofluorescence and skin elasticity are related to chronological age and body mass index of healthy subjects. Exp Gerontol. 2008;43(7):663–7. Epub 2008 Feb 9.

Darvin ME, Patzelt A, et al. One-year study on the variation of carotenoid antioxidant substances in living human skin: influence of dietary supplementation and stress factors. J Biomed Opt. 2008;13(4):044028.

Darvin ME, Fluhr JW, et al. In vivo distribution of carotenoids in different anatomical locations of human skin: comparative assessment with two different Raman spectroscopy methods. Exp Dermatol. 2009;18(12):1060–3.

Doukas AG, Soukos NS, et al. Fluorescence excitation spectroscopy for the measurement of epidermal proliferation. Photochem Photobiol. 2001;74(1):96–102.

Dyer DG, Dunn JA, et al. Accumulation of Maillard reaction products in skin collagen in diabetes and aging. J Clin Invest. 1993;91(6):2463–9.

Ediger MN, Olson BP, et al. Noninvasive optical screening for diabetes. J Diabetes Sci Technol. 2009;3(4):776–80.

Edwards EA, Duntley SQ. The pigments and color of human skin. Am J Anat. 1939;65:1–33.

Ermakov IV, Ermakova MR, et al. Noninvasive selective detection of lycopene and beta-carotene in human skin using Raman spectroscopy. J Biomed Opt. 2004;9(2):332–8.

Ermakov IV, Sharifzadeh M, et al. Resonance Raman detection of carotenoid antioxidants in living human tissue. J Biomed Opt. 2005;10(6):064028.

Gerstbrein B, Stamatas G, et al. In vivo spectrofluorimetry reveals endogenous biomarkers that report healthspan and dietary restriction in *Caenorhabditis elegans*. Aging Cell. 2005; 4(3):127–37.

Gillbro T, Cogdell RJ. Carotenoid fluorescence. Chem Phys Lett. 1989;158(3–4):312–6.

Gillies R, Zonios G, et al. Fluorescence excitation spectroscopy provides information about human skin in vivo. J Invest Dermatol. 2000;115(4):704–7.

Grether GF, Cummings ME, et al. Countergradient variation in the sexual coloration of guppies (*Poecilia reticulata*): drosopterin synthesis balances carotenoid availability. Evolution. 2005;59(1):175–88.

Groner W, Winkelman JW, et al. Orthogonal polarization spectral imaging: a new method for study of the microcirculation. Nat Med. 1999;5(10):1209–12.

Hagens R, Khabiri F, et al. Non-invasive monitoring of oxidative skin stress by ultraweak photon emission measurement. II: biological validation on ultraviolet A-stressed skin. Skin Res Technol. 2008;14(1):112–20.

Hammond Jr BR, Wooten BR. Validity issues with the in vivo measurement of skin carotenoids using Raman spectroscopy. J Invest Dermatol. 2004;122(2):544–6; author reply 546–8.

Hammond BR, Wooten BR. Resonance Raman spectroscopic measurement of carotenoids in the skin and retina. J Biomed Opt. 2005;10(5):054002.

Hartog JW, de Vries AP, et al. Accumulation of advanced glycation end products, measured as skin autofluorescence, in renal disease. Ann N Y Acad Sci. 2005;1043:299–307.

Hata TR, Scholz TA, et al. Non-invasive raman spectroscopic detection of carotenoids in human skin. J Invest Dermatol. 2000;115(3):441–8.

Hesterberg K, Lademann J, et al. Raman spectroscopic analysis of the increase of the carotenoid antioxidant concentration in human skin after a 1-week diet with ecological eggs. J Biomed Opt. 2009;14(2):024039.

Hull E, Ediger M, et al. Noninvasive, optical detection of diabetes: model studies with porcine skin. Opt Express. 2004;12(19):4496–510.

Kollias N, Baqer AH. Quantitative assessment of UV-induced pigmentation and erythema. Photodermatol. 1988;5(1):53–60.

Kollias N, Stamatas GN. Optical non-invasive approaches to diagnosis of skin diseases. J Investig Dermatol Symp Proc. 2002;7(1):64–75.

Kollias N, Gillies R, et al. A single parameter, oxygenated hemoglobin, can be used to quantify experimental irritant-induced inflammation. J Invest Dermatol. 1995;104(3):421–4.

Kollias N, Gillies R, et al. Fluorescence photography in the evaluation of hyperpigmentation in photodamaged skin. J Am Acad Dermatol. 1997;36(2 Pt 1):226–30.

Kollias N, Stamatas GN, et al. Suppression of UVB-induced cutaneous erythema by a previous UVB exposure. Photochem Photobiol. 2001;74(3):471–6.

Lingelbach LB, Mitchell AE, et al. Accumulation of advanced glycation endproducts in aging male Fischer 344 rats during long-term feeding of various dietary carbohydrates. J Nutr. 2000;130(5):1247–55.

Lopez-Sanchez M, Ayora-Canada MJ, et al. Olive fruit growth and ripening as seen by vibrational spectroscopy. J Agric Food Chem. 2010;58(1):82–7.

Lutgers HL, Graaff R, et al. Skin autofluorescence as a noninvasive marker of vascular damage in patients with type 2 diabetes. Diabetes Care. 2006;29(12):2654–9.

Meerwaldt R, Hartog JW, et al. Skin autofluorescence, a measure of cumulative metabolic stress and advanced glycation end products, predicts mortality in hemodialysis patients. J Am Soc Nephrol. 2005;16(12):3687–93. Epub 2005 Nov 9.

Meerwaldt R, Lutgers HL, et al. Skin autofluorescence is a strong predictor of cardiac mortality in diabetes. Diabetes Care. 2007;30(1):107–12.

Monami M, Lamanna C, et al. Skin autofluorescence in type 2 diabetes: beyond blood glucose. Diabetes Res Clin Pract. 2008;79(1):56–60. Epub 2007 Aug 20.

Monnier VM, Sell DR, et al. Glycation products as markers and predictors of the progression of diabetic complications. Ann N Y Acad Sci. 2005;1043:567–81.

Nandhini TA, Thirunavukkarasu V, et al. Taurine prevents fructose-diet induced collagen abnormalities in rat skin. J Diabetes Complications. 2005;19(5):305–11.

Novelli M, Masiello P, et al. Protein glycation in the aging male Sprague-Dawley rat: effects of antiaging diet restrictions. J Gerontol A Biol Sci Med Sci. 1998;53(2):B94–101.

Odetti PR, Borgoglio A, et al. Age-related increase of collagen fluorescence in human subcutaneous tissue. Metabolism. 1992;41(6):655–8.

Ou-Yang H, Stamatas G, et al. A chemiluminescence study of UVA-induced oxidative stress in human skin in vivo. J Invest Dermatol. 2004;122(4):1020–9.

Phillips SB, Kollias N, et al. Polarized light photography enhances visualization of inflammatory lesions of acne vulgaris. J Am Acad Dermatol. 1997;37(6):948–52.

Prince MR, Frisoli JK. Beta-carotene accumulation in serum and skin. Am J Clin Nutr. 1993;57(2):175–81.

Pudney PD, Melot M, et al. An in vivo confocal Raman study of the delivery of trans retinol to the skin. Appl Spectrosc. 2007;61(8):804–11.

Rerksuppaphol S, Rerksuppaphol L. Effect of fruit and vegetable intake on skin carotenoid detected by non-invasive Raman spectroscopy. J Med Assoc Thai. 2006;89(8):1206–12.

Rizwan M, Rodriguez-Blanco I, et al. Tomato paste rich in lycopene protects against cutaneous photodamage in humans in vivo. Br J Dermatol. 2010;21(10):1365–2133.

Ruiz D, Reich M, et al. Application of reflectance colorimeter measurements and infrared spectroscopy methods to rapid and nondestructive evaluation of carotenoids content in apricot (*Prunus armeniaca* L.). J Agric Food Chem. 2008;56(13):4916–22. Epub 2008 Jun 17.

Rutter K, Sell DR, et al. Green tea extract suppresses the age-related increase in collagen crosslinking and fluorescent products in C57BL/6 mice. Int J Vitam Nutr Res. 2003;73(6):453–60.

Sajithlal GB, Chithra P, et al. Effect of curcumin on the advanced glycation and cross-linking of collagen in diabetic rats. Biochem Pharmacol. 1998;56(12):1607–14.

Salares VR, Young NM, et al. Resonance Raman spectra of lobster shell carotenoproteins and a model astaxanthin aggregate. A possible photobiological function for the yellow protein. Biochemistry. 1977;16(21):4751–6.

Sauermann G, Mei WP, et al. Ultraweak photon emission of human skin in vivo: influence of topically applied antioxidants on human skin. Methods Enzymol. 1999;300:419–28.

Sprigle S, Zhang L, et al. Detection of skin erythema in darkly pigmented skin using multispectral images. Adv Skin Wound Care. 2009;22(4):172–9.
Stahl W, Heinrich U, et al. Increased dermal carotenoid levels assessed by noninvasive reflection spectrophotometry correlate with serum levels in women ingesting Betatene. J Nutr. 1998;128(5):903–7.
Stamatas GN, Kollias N. In vivo documentation of cutaneous inflammation using spectral imaging. J Biomed Opt. 2007;12(5):051603.
Stamatas GN, Estanislao RB, et al. Facial skin fluorescence as a marker of the skin's response to chronic environmental insults and its dependence on age. Br J Dermatol. 2006a;154(1):125–32.
Stamatas GN, Southall M, et al. In vivo monitoring of cutaneous edema using spectral imaging in the visible and near infrared. J Invest Dermatol. 2006b;126(8):1753–60. Epub 2006 May 4.
Stamatas GN, de Sterke J, et al. Lipid uptake and skin occlusion following topical application of oils on adult and infant skin. J Dermatol Sci. 2008a;50(2):135–42. Epub 2008 Mar 4.
Stamatas GN, Zmudzka BZ, et al. In vivo measurement of skin erythema and pigmentation: new means of implementation of diffuse reflectance spectroscopy with a commercial instrument. Br J Dermatol. 2008b;159(3):683–90. Epub 2008 May 28.
Stirban A, Nandrean S, et al. Skin autofluorescence increases postprandially in human subjects. Diabetes Technol Ther. 2008;10(3):200–5.
Thirunavukkarasu V, Nandhini AT, et al. Fructose diet-induced skin collagen abnormalities are prevented by lipoic acid. Exp Diabesity Res. 2004;5(4):237–44.
Tian WD, Gillies R, et al. Aging and effects of ultraviolet A exposure may be quantified by fluorescence excitation spectroscopy in vivo. J Invest Dermatol. 2001;116(6):840–5.
Werfel T. Skin manifestations in food allergy. Allergy. 2001;56 Suppl 67:98–101.
Yoshida S, Zhang QZ, et al. Metabolism of fatty acids and lipid hydroperoxides in human body monitoring with Fourier transform Infrared Spectroscopy. Lipids Health Dis. 2009;8:28.
Zidichouski JA, Mastaloudis A, et al. Clinical validation of a noninvasive, Raman spectroscopic method to assess carotenoid nutritional status in humans. J Am Coll Nutr. 2009;28(6):687–93.
Zonios G, Bykowski J, et al. Skin melanin, hemoglobin, and light scattering properties can be quantitatively assessed in vivo using diffuse reflectance spectroscopy. J Invest Dermatol. 2001;117(6):1452–7.

Chapter 14
Nutritional Clinical Studies in Dermatology

Aikaterini I. Liakou*, Michael J. Theodorakis*, and Christos C. Zouboulis

Core Messages

- Nutrition has long been associated with playing a principal and multifaceted role in skin health, beauty, integrity, and aging, both directly or indirectly, through multiple pathways and cofactors implicated in skin biology.
- The onset and the clinical courses of various common skin diseases, including acne, psoriasis, atopic dermatitis, and hair loss, have been shown to be critically affected by nutritional patterns and habits.
- Abnormal nutritional conditions, such as obesity anorexia nervosa, manifest in specific cutaneous appearance features and altered function of the skin.
- Skin photoprotection, rendered by various nutrients, has been well documented.
- Appropriate nutritional supplementation has been shown to exert beneficial effects on impaired skin's structural and functional integrity, restore its appearance, and promote skin health.

14.1 Introduction

The structural integrity, functional capacity, and regenerative potential of human skin are known to be influenced, to a variable extent, by a plethora of factors that greatly affect our appreciation of its overall appearance and our perception of its health and beauty. Heredity, sunlight, environmental or occupational exposure,

*Both authors contributed equally to the manuscript.

C.C. Zouboulis (✉)
Departments of Dermatology, Venereology, Allergology, and Immunology,
Dessau Medical Center, Auenweg 38, 06847 Dessau, Germany
e-mail: christos.zouboulis@klinikum-dessau.de

A. Pappas (ed.), *Nutrition and Skin: Lessons for Anti-Aging, Beauty and Healthy Skin*,
DOI 10.1007/978-1-4419-7967-4_14, © Springer Science+Business Media, LLC 2011

chronic disease, medications, drug abuse, hormonal supplementation, psychosomatic stress, and poor socioeconomic conditions have all been implicated in the pathophysiology of skin abnormalities and aging. Nutrition has historically been one of the earliest and most important factors strongly associated with skin health. However, the degree of its impact on skin physiology and the mechanisms involved in nutrition-dependent alterations in skin structure and function remain highly controversial. The practical difficulty of explicitly establishing a causative association between consumption of specific nutrients or food products and their potential effects on the skin have been hampered in clinical research, even when the relationship appears to be straightforward or highly reasonable. In this chapter, we attempt to analyze and review available evidence from the most important clinical nutritional studies in regard to human skin and provide scientifically sound and clinically relevant conclusions based on the emerging knowledge in this challenging area of research.

14.2 Effects of Nutrition on Specific Diseases in Dermatology

14.2.1 Nutrition and Acne

Acne has traditionally been a skin disease considered to be strongly associated with nutritional behavioral patterns and habits. However, based on a comprehensive review of the literature, up until 2005 there was no conclusive evidence of any true effects of diet on acne (Cordain 2005). Within the dermatology medical community, a consensus seems to have emerged that diet is indeed unrelated to the etiology or progression of acne. This finding might sound surprising, but one needs to consider the fact that theoretical conclusions on the issue were derived from a limited number of poorly designed studies, in some cases more than three decades old, which only randomly contained objective data. In addition, early studies exhibited many methodological limitations: small sample size, lack of appropriate controls, potential recall bias, incompletely reported results, or failure to define the clinical changes in acne (Anderson 1971; Fulton et al. 1969).

Interestingly, some studies have linked acne to milk consumption (Adebamowo et al. 2005). The investigators raised the point that most of the milk and dairy products consumed in the United States (where the study was conducted) came from pregnant cows. Could they have been responsible for acne pathogenesis via exposure to milk rich in hormones and other substances expressed during pregnancy? This hypothesis might be valid because sebum production is known to be influenced by androgens and hormonal mediators, such as sex hormone binding globulin (SHBG) and insulin-like growth factor-1 (IGF-1), each of which may be also influenced by dietary factors. The study was based on a questionnaire given to a group of 47,355 women who were asked to remember what they ate in high school, years prior to the study (Pappas 2009).

In a later study, teenage boys were asked to recall what they ate and to self-determine their acne severity (Adebamowo et al. 2008). Researchers concluded that there was an association between drinking milk and acne. However, all those studies had inherent limitations as the questionnaire required self-assessment of acne and was based on accurate memory of food intake. Factors such as heredity were ignored, and the data revealed a very low prevalence rate of acne. Furthermore there was a 20% increase in the prevalence of acne in milk drinkers in that study, based solely on memory. Another confusing aspect of that study is that a reverse association was reported between the consumption of milk fat and acne. Most of the hormones present in milk, especially the steroids, partition with the milk fat the same way that fat-soluble vitamins do. Hence, this result is bewildering as no matter how many hormones are left behind in the skim milk the whole milk should have a higher concentration of steroids (Pappas 2009). Evidently, these studies demonstrated a positive association between milk intake and acne. The low prevalence rates, memory test, self-assessment, and hormone speculation were, however, not significant enough to drive a recommendation or report an association of acne to milk. An important point is that dermatologists should not ignore the vast amount of literature on the reverse association of milk or calcium to obesity (Elwood 2005; Zemel and Sun 2008; Zemel 2005; Pereira et al. 2002). Even if milk is responsible for elevated insulin levels, is noteworthy that higher dairy intake, especially low-fat dairy intake, may lower the risk of type 2 diabetes in men and women (Choi et al. 2005; Liu et al. 2006).

It would be great to avoid having a cocktail of hormones in our daily diet, but we could not assume that each person in our society has access or can afford organic or hormone-free milk. Indeed, insulin and a high glycemic index are perhaps the two factors most scientifically and clinically associated with acne. Soon after the guidelines of the American Academy of Dermatology were published (Strauss et al. 2007), two clinical studies reported an association between high-glycemic-load diet and acne. Smith et al. (2007a, b, 2008), focused on the glycemic load, insulin sensitivity, hormonal mediators, and acne. The investigators reported that foods with a high glycemic index may contribute to acne by elevating serum insulin concentrations (thus stimulating sebocyte proliferation and sebum production), suppressing SHBG and raising androgen levels.

14.2.2 Nutrition and Atopic Disease

Development, clinical sequelae, and remission/relapse cycles of atopic disease (AD) have been attributed to food macro- and micronutrients (Helm 2004). Specific nutritional interventions using probiotics have been described in some studies to exert beneficial effects on the treatment and/or prevention of AD, with a decrease in the SCORAD score or a decrease of the frequency of AD during the first 2 years of life (Betsi et al. 2008). In a recent report, Koch et al. (2008) have shown the beneficial effect (decrease of SCORAD) of docohexaenoic acid supplementation in atopic eczema. In a similar fashion, oral evening primerose oil was reportedly of benefit to patients with moderate to severe eczema (Wright and Burton 1982), whereas a dietary supplement with fish oil has been shown to improve clinical management of psoriasis and eczema (Burton 1989).

14.2.3 Nutrition and Psoriasis

Diet has been suggested to participate in the etiology and pathogenesis of psoriasis, a T-cell-mediated inflammatory disease characterized by hyperproliferation and poor differentiation of epidermal keratinocytes. Protracted fasting and low-energy and vegetarian diets were shown to improve psoriasis symptoms in some studies, as were diets rich in ω-3 polyunsaturated fatty acids from fish oil, possibly by acting in a "antiinflammatory" fashion and thus by modifying the polyunsaturated fatty acid metabolism and eicosanoid profile. On the other hand, it has also been established that severe psoriasis may lead to nutrient depletion, especially of protein, folate, and iron. Nutrient deficit events have been attributed mainly to accelerated loss from the hyperproliferation and desquamation of the epidermal layer of skin in psoriasis, so this should be taken into account when impaired nutrition aggravates disease manifestations or predisposes to skin co-morbidities (Wolters 2005; Prystowsky et al. 1993).

The role of selenium has long been underscored in regard to duration and severity of psoriasis. However, such an association appears to exist in patients with long-standing psoriasis (>3 years) (Kharaeva et al. 2009; Serwin et al. 2003, 2006; Pinton et al. 1995).

In several studies dietary fish oil has been found to have beneficial effect on psoriasis, but dietary supplementation with very-long-chain ω-3 fatty acids was no better than corn-oil supplementation in treating psoriasis (Søyland et al. 1993).

14.2.4 Nutrition and Hair Loss

Hair loss is a common problem for both sexes, affecting up to 80% of men and 50% of women during their lifetime (Tosti et al. 2005; Rushton 2002). Nutrition and caloric restriction, as for example by malnutrition syndromes or anorexia nervosa, are well known to affect hair health and loss. Nutrients rich in antioxidants, primarily polyunsaturated fatty acids, zinc, taurine, and plant polyphenols, have been shown to restore a more balanced hair cycle, leading to decreased hair loss and increased hair density together with improved hair quality (Benyacoub et al. 2008).

14.2.5 Nutrition and Other Skin Diseases

In recent decades, the incidence of subjects presenting with reactive skin has considerably increased in industrial countries. Reactive skin is characterized by marked sensitivity of the skin to physical (heat, cold, wind) or chemical (topical product application) stimuli and occasionally by impaired ability for the skin barrier function to recover. One study demonstrated that after 43 days of supplementation a specific probiotic called *Lactobacillus paracasei* significantly decreased skin sensitivity compared to placebo and has also increased the recovery rate of the skin barrier function induced by mechanical disruption (Stahl et al. 2001).

14.2.6 Nutrition and Photoprotection

The most frequent damage induced by ultraviolet (UV) exposure is sunburn, and there has been evidence of its prevention by nutritional supplementation. β-Carotene (15–180 mg/day) and lycopene (up to 10 mg/day), two efficient oxygen quenchers, have been shown to prevent sunburn in humans (Stahl et al. 2006; Sies and Stahl 2004; Köpcke and Krutmann 2008). Systemic administration of antioxidants such as vitamin C (2 mg/day) and vitamin E (1,000 IU/day) as well as dietary fish oil (2 g/day) rich in ω-3 free fatty acids increased the minimum erythema dose (MED) (Eberlein-König et al. 1998; Rhodes et al. 1994). The effect of fish oil on UV-induced inflammation may be partially explained by its ability to reduce prostaglandin E_2 levels (Rhodes et al. 1995). Polyphenols provided by ingestion of cocoa rich in high flavanol (326 mg/day) reduced UV-induced erythema (Heinrich et al. 2006). *Polypodium leucotomos* (7.5 mg/kg body weight), a tropical fern plant used traditionally in Central America for the treatment of antiinflammatory disorders, has also been shown to counteract the erythematogenic effect of UV exposure (Middelkamp-Hup et al. 2004). Finally a specific association of a probiotic (*Lactobacillus johnsonii*, La1) with carotenoids (β-carotene 4.8 mg/day; lycopene 2 mg) was also shown to increase the MED (Bouilly-Gauthier et al. 2008).

Ultraviolet exposure can lead to both direct and indirect DNA damage. The major direct DNA damage is the release of cyclobutane pyrimidine dimers (thymine dimers and 6–4 photoproducts). Placzek et al. (2005) have shown that oral administration of vitamin C (2 mg/day) and vitamin E (1,000 IU/day) over 3 months had a protective effect against UV-induced thymine dimers.

Photoprotection by nutrients is well documented. Skin exposure to UV radiation leads directly or indirectly, through the generation of reactive oxygen species, to a large range of photodamage affecting cellular lipids, proteins, and DNA. It is involved in erythema appearance, premature skin aging, photoimmunosuppression, and skin cancer (Matsumura and Ananthaswany 2004; Fisher et al. 2002; Schwarz 2002).

Exposure to UV rays also causes local and systemic immunosuppression. Several mechanisms are involved in this deleterious effect, among which is the UV-induced depletion of Langerhans cells, the major antigen-presenting cells in the skin (Schwarz 2005). A placebo-controlled study demonstrated that β-carotene (30 mg/day) protects against photoimmunosuppression (Fuller et al. 1992). Oral administration of *Polypodium leucotomos* (1,080 mg) prior to UV exposure seemed to protect CD1a + cell density and preserve the dendricity of immune cells (Gonzales et al. 1997).

More recently, oral supplementation with the probiotic strain *Lactobacillus johnsonii* has been shown to accelerate the recovery of human skin immune homeostasis after UV-induced damage. This specific strain associated with carotenoids (β-carotene 4.8 mg/day; lycopene 2 mg) was also able to counteract UV-induced decrease of Langerhans cell density in human volunteers (Bouilly-Gauthier et al. 2010).

There is ample emerging evidence that probiotic strains can modulate the immune system of the skin in a beneficial way, leading to preservation of skin homeostasis. This could enable the design of novel nutrition-based compounds and interventions for preventing UV-induced damaging effects (Guéniche et al. 2009).

14.3 Nutritional Abnormalities and the Skin

14.3.1 Malnutrition

Primary nutritional deficiencies might be considered rare, but they are still prevalent in developing countries and should also be considered in developed countries in the setting of genetically predisposed disease states (MacDonald and Forsyth 2005). Most malnutrition syndromes in current medical practice are related to secondary elementary or macronutrient deficits due to prematurity (infants) or are seen in patients with long-term total parenteral nutrition, gastrointestinal pathology such as Crohn's disease, neoplasias, cystic fibrosis, or intestinal bypass procedures. They are also seen in chronic alcoholics and in individuals on restrictive diets (Tuerk and Fazel 2009). The skin is commonly involved and is often one of the first organs affected in nutritional deficiency, thus providing a key to the diagnosis. The most commonly encountered nutritional deficit disorders relate to zinc, biotin, essential fatty acid, protein deficiency, and kwashiorkor [common in patients with a human immunodeficiency virus (HIV) infection] (Ryan and Goldsmith 1996).

14.3.2 Eating Disorders: Anorexia Nervosa and Bulimia Nervosa

Many authors have reported skin signs in anorexia nervosa (AN) and bulimia nervosa (BN) (Tyler et al. 2002; Strumia et al. 2001, 2003; Glorio et al. 2000a). Cutaneous manifestations constitute somatic expression of underlying disorders, vomiting, abuse of drugs such as laxatives and diuretics, and psychiatric morbidity. Gupta et al. (1987) classified skin manifestations of eating disorders into four groups: those due to (1) starvation and/or malnutrition; (2) self-vomiting; (3) drug consumption; or (4) concomitant psychiatric illness. Glorio et al. (2000b) further identified two main groups of signs: (1) frequent signs (xerosis, alopecia, caries, opaque and fragile hair, nail fragility); and (2) guiding signs (hypertrichosis, Russell's sign, perimylolysis, self-induced dermatitis). Hediger et al. (2000) documented that a body mass index (BMI) of ≤16 should be considered a threshold value at or beyond which skin changes are more frequent.

Symptoms due to starvation include, in order of frequency: xerosis, lanugo-like body hair, telogen effluvium, carotenoderma, acne, hyperpigmentation, seborrheic dermatitis, acrocyanosis, perniosis, petechiae, livedo reticularis, interdigital intertrigo, paronychia, generalized pruritus, acquired striae distensae, slow wound healing, prurigo, edema, linear eczema craquelé, acral coldness, pellagra and scurvy, acrodermatitis entheropathica (Mitchell and Crow 2006).

Lanugo-like body hair is a frequent sign of AN, especially in young patients. It presents as fine, downy, pigmented hairs on the back, abdomen, and forearms. It is not a sign of virilization and has been associated with decreased activity of the 5α-reductase enzyme system, probably due to hypothyroidism.

Acne might be referred to starvation but could itself be a risk factor for AN. In fact, in psychologically vulnerable girls, a new diet behavior, adopted to control their acne, might actually lead to weight loss and AN. Moreover, the prevalence of acne is difficult to evaluate owing to age, which naturally predisposes to the disease. Carotenoderma is due to marked ingestion of carotenoid-rich vegetables low in calories. Acrocyanosis could represent a more extreme form of a heat-conserving mechanism not uncommon in anorectics. Raynaud's phenomenon and perniosis due to endocrinological complications have been also reported.

Purpura is the result of bone marrow depression due to starvation and subsequent thrombocytopenia. Life-threatening episodes of thrombocytopenia are reported in the typical restricting-type of AN with purpura, gingival, nasal and gastrointestinal bleeding, and apparent bone marrow hypoplasia (Abella et al. 2002). Nail fragility, longitudinal ungueal striae, onychocryptosis, periungueal erythema, prurigo pigmentosa, pompholyx, eruptive neurofibromatosis, evident blood vessels due to decreased subcutaneous tissue and acquired pili torti have also been reported.

The most characteristic cutaneous sign of purging-type AN is Russell's sign (knuckle calluses). The lesions involve calluses on the dorsal aspects of the dominant hand induced by the patients' repeated introduction of the hand into the mouth. It is a guiding sign in the diagnosis of eating disorders. With purging-type AN, patients may experience adverse reactions of drugs, such as laxatives, diuretics, and appetite suppressants, which they often use. Self-induced trauma often coexists with AN, varying from unconscious picking at the skin to severe self-destructive actions (Strumia 2009).

14.3.3 Obesity

Obesity has been associated with multiple skin disorders, including altered skin barrier function, sebaceous gland physiology and sebum production, sweat gland biology and regulation, lymphatic drainage, collagen structure and functional properties, process of wound healing, distribution and pathobiology of subcutaneous adipose tissue, and impaired microcirculatory supply (Yosipovitch et al. 2007). Characteristic dermatological signs with obesity—particularly pronounced in morbid obesity and when co-morbidities such as the polycystic ovaries syndrome are present—include acanthosis nigricans, acrochordons, keratosis pilaris, hyperandrogenism, hirsutism, striae distensae, adiposis dolorosa, fat redistribution, lymphedema, chronic venous insufficiency, plantar hyperkeratosis, cellulitis, skin infections, hidradenitis suppurativa, tophaceous gout, and eruption of psoriasis and various dermatoses (Jabbour 2003; Krause 2008).

14.4 Nutrition, Skin Aging, and Skin Beauty

Increased life expectancy is associated with a need to appear healthy and handsome. Many attempts have been made to improve skin health and beauty by changing or supplementing the diet. In 2001, Boelsma et al. (2001) reviewed the effects of vitamin, carotenoid, and fatty acid supplementation for optimizing the skin condition and preventing skin diseases. They concluded that nutritional factors show potential beneficial actions on the skin. Recently, epidemiological evidence suggested that multivitamin use is associated with longer telomere length, a marker of biological aging, in women (Xu et al. 2009).

Skin, especially facial skin, is one of the most important factors in attractivity; therefore, prevention of premature signs of skin aging are considered among the top priorities. Two reports have provided evidence that food and nutrient intake can indeed influence skin aging. In 2001, Purba et al. (2001) noted that actinic damage, especially skin wrinkling, may be associated with poor food habits. In that study, high intake of vegetables, legumes, and olive oil appeared to be protective against cutaneous actinic damage. In another report, higher intake of vitamin C and linoleic acid, as well as lower intake of fat and carbohydrate, were shown to be associated with better skin appearance (Cosgrove et al. 2007). In addition to the traditional use of topical care, nutritional supplements have emerged as a new strategy to improve skin beauty.

Yamakoshi et al. (2004) investigated the effect of oral intake of a proanthocyanidine-enriched extract (201 mg/day over 6 months) on facial hyperpigmentation in women and demonstrated that this extract was able to improve chloasma, determined by clinical evaluation as well as by a colorimetric method. Oral fish polysaccharides (3 × 250 mg day over 8 weeks) associated with an antioxidant mix have been shown to improve dermal thickness, skin wrinkling, color, and viscoelasticity after 2 months of supplementation (Distante et al. 2002).

Silicon (20 mg/day for 20 weeks) enhanced skin microrelief and mechanical properties in women with photo-damaged skin (Barel et al. 2005). A combination of lycopene (6 mg), vitamin C (60 mg), and soy isoflavones (50 mg) has been shown to maintain skin density, improve skin firmness, and provide microrelief, hydration, and tone in menopausal women (Dréno 2003; Piccardi and Manissier 2009; Zouboulis et al. 2009).

14.5 Conclusion

For a long time nutrition has been considered one of the principal factors influencing overall "well-being" and the perception of "health" in humans. The skin, being the largest and heaviest endocrine organ in the body (Zouboulis 2009), provides a first impression about one's biological condition, age, and beauty. It reflects the psychosomatic balance and stress status. It also constitutes a prime target for ingested nutrients, either directly or indirectly. Increasing clinical appreciation of the growing significance of nutritional composition, patterns, habitual exposure and its interplay with multiple hormonal mediators, pathway cofactors, structural elements, and functional parameters of the skin have been underscored by emerging evidence from the literature. Not only have nutritional deficiencies or excesses been shown to predispose to the onset or recurrence of various dermatological disorders, they are involved in the pathogenesis and clinical manifestations. On the other hand, accurate knowledge and appropriate handling of our potential to manipulate individual nutritional aspects as treatment modalities in skin pathologies could provide a powerful, patient-friendly tool to prevent, alleviate, or even cure common diseases in dermatology.

Take-Home Messages

- Proper nutrition in both caloric content and composition is reflected in healthy skin.
- Nutritional imbalances are mainly reflected in characteristic skin pathologies.
- Hyperinsulinemia and ingestion of high-glycemic-index foods are the two factors most frequently associated with acne.
- Xerosis, alopecia, caries, opaque and fragile hair, and nail fragility are frequent cutaneous manifestations of anorexia nervosa.
- Probiotics preserve skin metabolism, having a positive effect in atopic disease and psoriasis.
- Antioxidants with polyunsaturated fatty acids, zinc, taurine, and plant polyphenols restore a more balanced hair cycle, leading to decreased hair loss.
- β-Carotene and lycopene have been shown to prevent sunburn in humans.
- Vitamins, carotenoids, and fatty acid supplementation have been proven to optimize skin physiology and prevent some skin diseases.
- Oral fish polysaccharides associated with an antioxidant mix have been shown to improve dermal thickness, skin wrinkling, color, and viscoelasticity.
- Lycopene, vitamin C, and soy isoflavones has been shown to maintain skin density, improve skin firmness, and provide microrelief, hydration, and tone in menopausal women.

References

Abella E, Feliu E, Granada I, Millá F, Oriol A, Ribera JM, et al. Bone marrow changes in anorexia nervosa are correlated with the amount of weight loss and not with other clinical findings. Am J Clin Pathol. 2002;118:582–8.

Adebamowo CA, Spiegelman D, Danby FW, et al. High school dietary dairy intake and teenage acne. J Am Acad Dermatol. 2005;52:207–14.

Adebamowo CA, Spiegelman D, Berkey CS, et al. Milk consumption and acne in teenaged boys. J Am Acad Dermatol. 2008;58:787–93.

Anderson PC. Foods as the cause of acne. Am Fam Physician. 1971;3:102–3.

Barel A, Calomme M, Timchenko A, De Paepe K, Demeester N, Rogiers V, et al. Effect of oral intake of choline-stabilized orthosilicic acid on skin, nails and hair in women with photodamaged skin. Arch Dermatol Res. 2005;297:147–53.

Benyacoub J, Gueniche A, Bureau-Franz I, Castiel I. Probiotiques et peau. In: Roberfroid M, Coxam V, Delzenne N, editors. Aliments Fonctionnels. 2me édition. Paris: Lavoisier; 2008.

Betsi GI, Papadavid E, Falagas ME. Probiotics for the treatment or prevention of atopic dermatitis: a review of the evidence from randomized controlled trials. Am J Clin Dermatol. 2008;2:93–103.

Boelsma E, Hendriks HFJ, Roza L. Nutritional skin care: health effects of micronutrients and fatty acids. Am J Clin Nutr. 2001;73:853–64.

Bouilly-Gauthier D, Jeannes C, Dupont N, Piccardi N, Manissier P, Heinrich U, Tronnier H. A new nutritional supplementation is effective against hair loss and improves hair quality. (2008) JEADV;FP1003 [Abstract].

Bouilly-Gauthier D, Jeannes C, Maubert Y, Duteil L, Queille-Roussel C, Piccardi N, et al. Clinical evidence of skin benefits of a dietary supplement containing probiotic and carotenoids on UV-induced damage. Br J Dermatol. 2010;163:536–43.

Burton JL. Dietary fatty acids and inflammation in skin disease. Lancet. 1989;1:27–31.

Choi HK, Willett WC, Stampfer MJ, Rimm E, Hu FB. Dairy consumption and risk of type 2 diabetes mellitus in men: a prospective study. Arch Intern Med. 2005;165:997–1003.

Cordain L. Implications for the role of diet in acne. Semin Cutan Med Surg. 2005;24:84–91.

Cosgrove MC, Franco OH, Granger SP, Murray PG, Mayes AE. Dietary nutrient intakes and skin-aging appearance among middle-aged women. Am J Clin Nutr. 2007;86:1225–31.

Distante F, Scalise F, Rona C, Bonfigli A, Fluhr JW, Berardesca E. Oral fish cartilage polysaccharides in the treatment of photoageing: biophysical findings. Int J Cosmet Sci. 2002;24:81–7.

Dréno B. New assessment methods applied to a patented lacto-lycopene, soy isoflavones and vitamin C in the correction of skin aging. Les Nouvelles Dermatol. 2003;22:1–6.

Eberlein-König B, Placzek M, Przybilla B. Protective effect against sunburn of combined systemic ascorbic acid (vitamin C) and d-alpha-tocopherol (vitamin E). J Am Acad Dermatol. 1998;38:45–8.

Elwood PC. Time to value milk. Int J Epidemiol. 2005;34:1160–2.

Fisher GJ, Kang S, Varani J, Bata-Csorgo Z, Wan Y, Datta S, et al. Mechanism of photoaging and chronological skin aging. Arch Dermatol. 2002;138:1462–70.

Fuller J, Faulkner H, Bendich A, Parker RS, Roe DA. Effect of β-carotene supplementation on photosuppression of delayed-type hypersensitivity in normal young men. Am J Clin Nutr. 1992;56:684–90.

Fulton Jr JE, Plewig G, Kligman AM. Effect of chocolate on acne vulgaris. JAMA. 1969;210:2071–4.

Glorio R, Allevato M, De Pablo A, et al. Prevalence of cutaneous manifestations in 200 patients with eating disorders. Int J Dermatol. 2000a;39:348–53.

Glorio R, Allevato M, De Pablo A, Abbruzzese M, Carmona L, Savarin M, et al. Prevalence of cutaneous manifestations in 200 patients with eating disorders. Int J Dermatol. 2000b;39:348–53.

Gonzales S, Pathak MA, Cuevas J, Villarrubia VG, Fitzpatrick TB. Topical or oral administration with an extract of *Polypodium leucotomos* prevents acute sunburn and psoralen-induced phototoxic reactions as weel as depletion of Langerhans cells in human skin. Photodermatol Photoimmunol Photomed. 1997;13:50–60.

Guéniche A, Philippe D, Bastien P, Blum S, Buyukpamukcu E, Castiel-Higounenc I. Probiotoics for photoprotection. Dermatoendocrinol. 2009;1:1–5.
Gupta MA, Gupta AK, Haberman HF. Dermatologic signs in anorexia nervosa and bulimia nervosa. Arch Dermatol. 1987;123:1386–90.
Hediger C, Rost B, Itin P. Cutaneous manifestations in anorexia nervosa. Schweiz Med Wochenschr. 2000;130:565–75.
Heinrich U, Neukam K, Tronnier H, Sies H, Stahl W. Long-term ingestion of high flavanol cocoa provides photoprotection against UV-induced erythema and improves skin condition in women. J Nutr. 2006;136:1565–9.
Helm RM. Diet and the development of atopic disease. Curr Opin Allergy Clin Immunol. 2004;4:125–9.
Jabbour SA. Cutaneous manifestations of endocrine disorders: a guide for dermatologists. Am J Clin Dermatol. 2003;4:315–31.
Kharaeva Z, Gostova E, De Luca C, Raskovic D, Korkina L. Clinical and biochemical effects of coenzyme Q(10), vitamin E, and selenium supplementation to psoriasis patients. Nutrition. 2009;25:295–302.
Koch C, Dölle S, Metzger M, Rashe C, Jungclas H, Rühl R, et al. Docohexaenoic acid (DHA) supplementation in atopic eczema: a randomized, double-blind, controlled trial. Br J Dermatol. 2008;158:786–92.
Köpcke W, Krutmann J. Protection from sunburn with β-carotene-A meta-analysis. Photochem Photobiol. 2008;84:284–8.
Krause W. Dermatoses in obesity: sick skin over the adipose tissue. MMW Fortschr Med. 2008;150:45–7.
Liu S, Choi HK, Ford E, Song Y, Klevak A, Buring JE, et al. A prospective study of dairy intake and the risk of type 2 diabetes in women. Diabetes Care. 2006;29:1579–84.
MacDonald A, Forsyth A. Nutritional deficiencies and the skin. Clin Exp Dermatol. 2005;30:388–90.
Matsumura Y, Ananthaswany HN. Toxic effects of ultraviolet radiation on the skin. Toxicol Appl Pharmacol. 2004;195:298–308.
Middelkamp-Hup MA, Pathak MA, Parrado C, Goukassian D, Rius-Díaz F, Mihm MC, et al. Oral *polypodium leucotomos* extract decreases ultraviolet-induced damage of human skin. J Am Acad Dermatol. 2004;51:910–8.
Mitchell JE, Crow S. Medical complications of anorexia nervosa and bulimia nervosa. Curr Opin Psychiatry. 2006;19:438–43.
Pappas A. The relationship of diet and acne. Dermatoendocrinol. 2009;1:262–7.
Pereira MA, Jacobs Jr DR, Van Horn L, Slattery ML, Kartashov AI, Ludwig DS. Dairy consumption, obesity, and the insulin resistance syndrome in young adults: the CARDIA Study. JAMA. 2002;287:2081–9.
Piccardi N, Manissier P. Nutrition and nutritional supplementation: impact on skin health and beauty. Dermatoendocrinol. 2009;1:271–4.
Pinton J, Fridén H, Kettaneh-Wold N, Wold S, Dreno B, Richard A, et al. Clinical and biological effects of balneotherapy with selenium-rich spa water in patients with psoriasis vulgaris. Br J Dermatol. 1995;133:344–7.
Placzek M, Gaube S, Kerkmann U, Gilbertz KP, Herzinger T, Haen E. Przybilla B. Ultraviolet B-induced DNA damage in human epidermis is modified by the antioxidants ascorbic acid and d-alpha-tocopherol. J Invest Dermatol. 2005;124:304–7.
Prystowsky JH, Orologa A, Taylor S. Update on nutrition and psoriasis. Int J Dermatol. 1993;32:582–6.
Purba M, Kouris-Blazos A, Wattanapenpaiboon N, Lukito W, Rothenberg EM, Steen BC, et al. Skin wrinkling: can food make a difference? J Am Coll Nutr. 2001;1:71–80.
Rhodes LE, O'Farrell S, Jackson MJ, Friedmann PS. Dietary fish-oil supplementation in humans reduces UVB-erythemal sensitivity but increase epidermal lipid peroxidation. J Invest Dermatol. 1994;103:151–4.
Rhodes LE, Durham BH, Fraser WD, Friedmann PS. Dietary fish oil reduces basal and ultraviolet B-generated PGE2 levels in skin and increases the threshold to provocation of polymorphic light eruption. J Invest Dermatol. 1995;105:532–5.

Rushton DH. Nutritional factors and hair loss. Clin Exp Dermatol. 2002;27:396–404.
Ryan AS, Goldsmith LA. Nutrition and the skin. Clin Dermatol. 1996;14:389–406.
Schwarz T. Photoimmunosuppression. Photodermatol Photoimmunol Photomed. 2002;18:141–5.
Schwarz T. Mechanisms of UV-induced immunosuppression. Keio J Med. 2005;54:165–71.
Serwin AB, Wasowicz W, Gromadzinska J, Chodynicka B. Selenium status in psoriasis and its relations to the duration and severity of the disease. Nutrition. 2003;19:301–4.
Serwin AB, Wasowicz W, Chodynicka B. Selenium supplementation, soluble tumor necrosis factor-alpha receptor type 1, and C-reactive protein during psoriasis therapy with narrowband ultraviolet B. Nutrition. 2006;22:860–4.
Sies H, Stahl W. Nutritional photoprotection against skin damage from sunlight. Annu Rev Nutr. 2004;24:173–200.
Smith RN, Mann NJ, Braue A, Mäkeläinen H, Varigos GA. A low-glycemic-load diet improves symptoms in acne vulgaris patients: a randomized controlled trial. Am J Clin Nutr. 2007a;86:107–15.
Smith RN, Mann NJ, Braue A, Mäkeläinen H, Varigos GA. The effect of a high-protein, low glycemic-load diet versus a conventional, high glycemic-load diet on biochemical parameters associated with acne vulgaris: a randomized, investigator-masked, controlled trial. J Am Acad Dermatol. 2007b;57:247–56.
Smith RN, Braue A, Varigos GA, Mann NJ. The effect of a low glycemic load diet on acne vulgaris and the fatty acid composition of skin surface triglycerides. J Dermatol Sci. 2008;50:41–52.
Søyland E, Funk J, Rajka G, Sandberg M, Thune P, Rustad L, et al. Effect of dietary supplementation with very-long-chain n-3 fatty acids in patients with psoriasis. N Engl J Med. 1993;328:1812–6.
Stahl W, Heinrich U, Wiseman S, Eicher O, Sies H, Tronnier H. Dietary tomato paste protects against ultraviolet light-induced erythema in humans. J Nutr. 2001;131:1449–51.
Stahl W, Heinrich U, Aust O, Tronnier H, Sies H. Lycopene-rich products and dietary photoprotection. Photochem Photobiol Sci. 2006;5:238–42.
Strauss JS, Krowchuk DP, Leyden JJ, Lucky AW, Shalita AR, Siegfried EC, et al. American Academy of Dermatology/American Academy of Dermatology Association. Guidelines of care for acne vulgaris management. J Am Acad Dermatol. 2007;56:651–63.
Strumia R. Skin signs in anorexia nervosa. Dermatoendocrinol. 2009;1:268–70.
Strumia R, Varotti E, Manzato E, Gualandi M. Skin signs in anorexia nervosa. Dermatology. 2001;203:314–7.
Strumia R, Manzato E, Gualandi M. Cutaneous manifestations in male anorexia nervosa; four cases. Acta Derm Venereol. 2003;83:464–5.
Tosti A, Piraccini M, Lorizzo M, Voudouris S. The natural history of androgenetic alopecia. J Cosmet Dermatol. 2005;4:41–3.
Tuerk MJ, Fazel N. Zinc deficiency. Curr Opin Gastroenterol. 2009;25:136–43.
Tyler I, Wiseman C, Crawford RI, Birmingham CL. Cutaneous manifestations of eating disorders. J Cutan Med Surg. 2002;6:345–53.
Wolters M. Diet and psoriasis: experimental data and clinical evidence. Br J Dermatol. 2005;153:706–14.
Wright S, Burton JL. Oral evening primerose sedd oil improves atopic eczema. Lancet. 1982;2:1120–2.
Xu Q, Parks CG, De Roo LA, Cawthon RM, Sandler DP, Chen H. Multivitamin use and telomere length in women. Am J Clin Nutr. 2009;89:1–7.
Yamakoshi J, Sano A, Tokutake S, Saito M, Kikuchi M, Kubota Y, et al. Oral intake of proanthocyanidin-rich extract from grape seeds improves chloasma. Phytother Res. 2004;18:895–9.
Yosipovitch G, DeVore A, Dawn A. Obesity and the skin: skin physiology and skin manifestations of obesity. J Am Acad Dermatol. 2007;56:901–16.
Zemel MB. The role of dairy foods in weight management. J Am Coll Nutr. 2005;24:537S–46.
Zemel MB, Sun X. Dietary calcium and dairy products modulate oxidative and inflammatory stress in mice and humans. J Nutr. 2008;138:1047–52.
Zouboulis CC. The skin as an endocrine organ. Dermatoendocrinol. 2009;1:1–3.
Zouboulis CC, Rabe T, Bayerl C. Sinn und Unsinn der ästhetischen Endokrinologie. Gynäkologische Endokrinologie. 2009;7:25–32.

Index

A. Pappas (ed.), *Nutrition and Skin: Lessons for Anti-Aging, Beauty and Healthy Skin*,
DOI 10.1007/978-1-4419-7967-4, © Springer Science+Business Media, LLC 2011